W9-CRB-687

Polarized light in Nature

Polarized light in Nature

G. P. KÖNNEN

Royal Netherlands Meteorological Institute (KNMI)
de Bilt, the Netherlands

TRANSLATED BY G. A. BEERLING

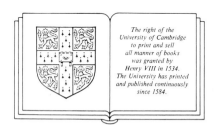

The right of the
University of Cambridge
to print and sell
all manner of books
was granted by
Henry VIII in 1534.
The University has printed
and published continuously
since 1584.

CAMBRIDGE UNIVERSITY PRESS

CAMBRIDGE
LONDON   NEW YORK   NEW ROCHELLE
MELBOURNE   SYDNEY

Published by the Press Syndicate of the University of Cambridge
The Pitt Building, Trumpington Street, Cambridge CB2 1RP
32 East 57th Street, New York, NY 10022, USA
10 Stamford Road, Oakleigh, Melbourne 3166, Australia

Originally published in Dutch as *Gepolariseerd licht in de Natuur* by
B.V. W.J. Thieme & Cie-Zutphen, Netherlands, 1980 and © B.V. W.J.
Thieme & Cie-Zutphen 1980
First published in English by Cambridge University Press 1985 as
Polarized light in Nature
English edition © Cambridge University Press 1985

Printed in the Netherlands by B.V. W.J. Thieme & Cie-Zutphen
Library of Congress catalogue card number: 83-18912

British Library Cataloguing in Publication Data

Können, G.P.
  Polarized light in Nature
  1. Polarization (Light)   2. Meteorological optics
  I. Title   II. Gepolariseerd licht in de
  Natuur. *English*
  551.5′6   QC976.P7

ISBN 0 521 25862 6

SE

Contents

Preface

Light is characterized by three properties: brightness, colour and polarization. The first two determine what impression we get from the world: the human eye is extremely sensitive to colour shades and differences in brightness. Polarization, on the other hand, is almost impossible to detect with the naked eye. This changes with the aid of a polarizing filter like those, for instance, in sunglasses; then all at once we perceive how much polarized light there really is. The world of polarized light differs amazingly from that seen without a filter: whereas some objects hardly change, others give us quite a new impression.

The subject of this book is two-fold. It is a guide to the colour and light phenomena around us and it focusses particularly on their polarization. Almost all the effects that are discussed can be seen through a simple polarizing filter without any further aids.

As far as I know, this is the first attempt to compile a guide to polarization: the literature on this subject is widespread and not always complete. This necessitated personal observations in the open field; indeed, about 90% of the effects under discussion have been observed by myself. This emphasizes how little attention has been paid previously to this subject and it is remarkable that there are really very striking polarization effects (particularly those of haloes) that apparently have never been described before.

The book has been arranged as follows. In Part I there is a general discussion on polarization and its observation. Part II is the proper guide to Nature, in which we 'arm' ourselves with a polarizing filter in order to see this aspect of the play of light in the sky, on the earth and in nocturnal surroundings. Special attention has been paid to the magnificent colourful optical phenomena in the sky which, in spite of their frequent appearances, are unfortunately only perceived by a few and whose polarization effects have been observed still less. This part ends with a survey of the colour patterns appearing in minerals lit by polarized light. Finally, in Part III, the formation of polarized light by various mechanisms is graphically described so that the reader can visualize which processes result in polarized light and which do not. Here I have consciously chosen the simplest possible approach. This part concludes with a general view of the polarization phenomena in Nature, explained on the basis of simple rules of symmetry.

Two type sizes are used in this book. The phenomenological description of the observations in Nature and the necessary background information on polarization have been printed in text-type; the practical reader may restrict himself to these parts. For the more specialist reader, more detailed background and some additional observations are included in small type. Consequently, Parts I and II are mainly in text-type and Part III is in small type.

I am especially indebted to my brother E. E. Können for the particularly scrupulous way in which he perused the original Dutch manuscript and for the many valuable suggestions resulting from his close attention. Moreover, I want to acknowledge Mrs J. Wevers for her help and encouragement with optical mineralogy and to all those who made their photographic material available to me or supported me otherwise. Last but not least, however, I would like to express my gratitude to my uncle, G. A. Beerling, for his work on the translation of the original Dutch version into English, and to thank Mrs S. Irvine for her advice and editorial work during the preparation of this English version.

G. P. Können

January 1984

Acknowledgements

Cover photographs
   Front, above: N. F. Verster
   Back: M. A. Posthumus
Photographs: G. P. Können, unless otherwise stated.
Drawings: G. Westerhof.
The polarizing filter in this book is kindly supplied by Polarizers Ltd, Oudegracht 90, Alkmaar 6.
Almost all polarized-light effects described in this book, are visible with this polarizing filter or with polaroid sunglasses.

Part I

Polarized light:
what it is and how to
observe it

Forms of polarized light

1. Polarized light and unpolarized light

Light is a *wave* of electromagnetic nature. Such a wave always vibrates normally to its direction of propagation (fig. 1). The wave in fig. 1 vibrates in the plane of this piece of paper. It need not always be so: for instance, such a wave can just as well vibrate perpendicularly to it or at a different angle. Sunlight, for example, has no special preference for the direction in which it vibrates, and the plane of vibration alternates many times per second. Such light is called *unpolarized* or *natural light*.

It turns out, however, that there is also light with vibrations remaining in one plane. Such light is called (*linearly*) *polarized*. This kind of light also occurs in Nature, and it is this particular form which is the subject of this book. The human eye is barely sensitive to the polarization of light, so that generally we have to manage with a filter to be able to see polarization effects. Indeed, the eye is as insensitive to polarization phenomena as it is sensitive to colour shades. By analogy with colour blindness, we can consider the human eye to be virtually '*polarization-blind*'. The more surprising, therefore, are the effects which can be observed in Nature with a polarizing filter such as polaroid sunglasses.

For the sake of convenience, we can consider light to be produced by a particle (e.g. an electron) which vibrates very fast to and fro, and

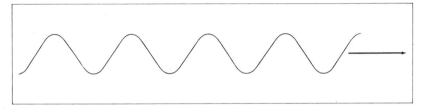

Fig. 1 A light wave vibrates perpendicularly to its direction of propagation.

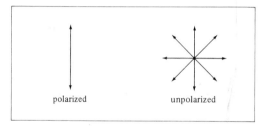

Fig. 2 Light vibrations of polarized and unpolarized (natural) light.

consequently emits a light wave. If this particle goes on vibrating in the same direction, it emits polarized light. When, however, it repeatedly vibrates in other directions, the polarization plane changes constantly; and unpolarized light is the result (fig. 2). Linearly polarized light is, therefore, characterized by *direction* (of vibration) in the same way as the colour of light is characterized by *length* (i.e. the wavelength of light waves). Polarization is not confined to light: it can occur in any electromagnetic radiation; thus radar waves, radio waves or X-rays can also be polarized.

It is important to note the following: *unpolarized* (natural) light can also be assumed to consist of only *two* vibrations of equal amplitudes, vibrating at right angles to each other. Mathematically this comes to the same thing as considering the sum of all possible vibrations. But the two vibrations must vibrate in randomly changing phase with respect to each other, so that they are not coherent. Such an approach makes it possible to indicate the state of polarization of light graphically with the aid of simple diagrams (see fig. 3).

Polarized light can also be decomposed into two vibrations, the directions of which are perpendicular to each other (see fig. 4). In this instance, however, they *do* vibrate coherently, i.e. the phase difference between them always remains the same (zero, in this case). Such analyses are important, when investigating what happens when the behaviour of light differs for a different direction of polarization.

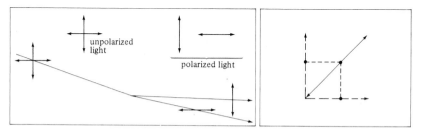

Fig. 3 Splitting of an unpolarized lightbeam into two polarized beams can be graphically shown by vibration diagrams as in fig. 2.

Fig. 4 Decomposition of polarized light into two light beams, vibrating perpendicularly to each other.

2. Linearly polarized light and circularly polarized light

Linear polarization is not the only form of polarized light. Besides vibrating to and fro, a particle can also make a *circular* motion. The light wave then emitted forms a kind of spiral. This is called *circular polarization*, which is not characterized by a vibration direction but *sense of rotation*. Left-handed or right-handed light can be distinguished, according to the counter-clockwise or clockwise motion of the particle (or wave) as viewed by the observer. Hence, the light wave illustrated in fig. 5 is right-handed.

It is important that circular light can also be conceived as the *sum* of two coherent linearly polarized waves which vibrate perpendicularly to each other. The *phases* of these two waves must now have a difference of 90°, i.e. one 'particle' must be in an extreme position, when the other is exactly between its extreme positions (fig. 6). Conversely, one can imagine linearly polarized light as consisting of two circular waves which rotate in opposite directions. Addition of these motions results in a linearly polarized wave.

Light consisting of a mixture of linearly and circularly polarized waves is the most general case of polarization. This is called *elliptically polarized light*, because the sum of the motions of two such vibrations (a circle and a line) forms an ellipse (fig. 7). One could say, therefore, that a circular component has been added to the linearly polarized light or vice versa. Elliptical light is, consequently, characterized by direction (as in the case of linearly polarized light) *and* by sense of rotation (as in the case of circularly polarized light). This light too can be considered as consisting of two linearly polarized waves, but the vibrations need no longer be perpendicularly directed to each other, and their phase-difference need not be 90°. Nor is it

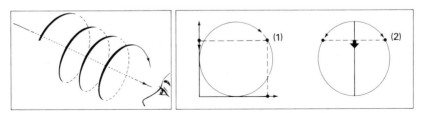

Fig. 5 Circularly polarized light wave (right-handed).

Fig. 6 (1) Circularly polarized light can be thought to be composed of two linearly polarized waves and (2) linearly polarized light composed of two circularly polarized waves.

Fig. 7 Elliptical light can be considered as the sum of linearly and circularly polarized light.

necessary that the amplitude of the vibrations should be identical. It is evident, indeed, that elliptical polarization is the most general form.

In this treatise, however, we shall imagine that this light is composed of a linear and a circular part – the simplest conception. So, we discuss only circularly and linearly polarized light and elliptically polarized light is considered merely as a *mixture* of the first two types.

3. Partially polarized light and degree of polarization

The light that reaches us is usually not completely polarized: such light is called *partially polarized*. In the case of linear polarization, this means that, although the vibration plane of the light waves has a preferential direction, vibrations in other planes are also permitted. In the most general case, partially polarized light has linearly polarized, circularly polarized and unpolarized components and the extent to which the light is polarized is called the *degree of polarization*. Diagrams for partial linear polarization are given in fig. 8.

The *degree of linear* polarization P is given by the intensity of light in the preferential direction of the plane of vibration ($I_\parallel$) minus the intensity of light perpendicular to it ($I_\perp$) and this divided by the total intensity of light which equals $I_\parallel + I_\perp$. Thus,

$$P = \frac{I_\parallel - I_\perp}{I_\parallel + I_\perp}$$

Usually, P is expressed in a percentage and hence ranges from 0% to 100%.

The *degree of circular* polarization q can be defined in a similar way and gives the fraction of circularly polarized light in the beam. When considering circular polarization alone, it is sensible to assume that unpolarized light is a mixture of two non-coherent circular waves of equal amplitude which are left-handed and right-handed. When one of the two dominates, however, the net effect is partially polarized circular light. Naturally, the *total* degree of polarization $(P^2 + q^2)^{\frac{1}{2}}$ must always be less than 100%.

Sometimes the term *negative* polarization is used. This means that the

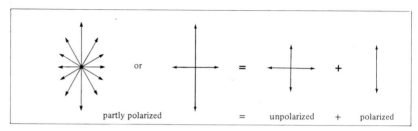

partly polarized = unpolarized + polarized

Fig. 8 Partially polarized light can be considered to consist of a totally polarized and a totally unpolarized part.

direction of polarization is changed: negative left-handed rotation corresponds to right-handed rotation. Negative linearly polarized light has its plane of polarization (i.e. the preferential direction of the vibration plane) perpendicular to a previously fixed plane.

There is, of course, a close relationship between the degree of linear polarization P, and $I_{||}$, and $I_{\perp}$. If for instance $P = 50\%$, then $I_{||}:I_{\perp} = 3:1$; so there is three times more light in the direction of the preferential plane than perpendicular to it. Other examples are given below.

P	$I_{||}:I_{\perp}$	P	$I_{||}:I_{\perp}$
5%	21 : 19 = 1.11	60%	4 : 1 = 4.0
10%	11 : 9 = 1.22	70%	17 : 3 = 5.7
20%	3 : 2 = 1.50	75%	7 : 1 = 7.0
30%	13 : 7 = 1.86	80%	9 : 1 = 9.0
40%	7 : 3 = 2.33	90%	19 : 1 = 19.0
50%	3 : 1 = 3.0	100%	1 : 0 = ∞

An exactly similar table can be drawn up for the degree of circular polarization q; here, however, we must read q instead of P, and $I_L:I_R$ or $I_R:I_L$ instead of $I_{||}:I_{\perp}$ (I_L and I_R are, respectively, the left-handed and right-handed intensities).

For a polarization degree of <5 at 10% light is spoken of as hardly polarized or non-polarized, between 5 at 10% and 15 at 20% it is slightly polarized, at 20–60% it is strongly polarized, and at $>60\%$ it is very strongly polarized. If the degree of polarization amounts to 100%, this is completely or totally polarized light.

There are processes in Nature that can convert linearly polarized, circularly polarized and unpolarized light into each other (reflection, for example). P or q are changed in this case. With linearly polarized light, the plane of polarization can be rotated, and for circularly polarized light the sense of rotation can be changed. In these cases P or q can remain the same, but the character of the polarized light changes. These mechanisms will be discussed in Part III and are of great importance in the origin and form of polarized light in Nature, although the greater part of the polarized light around us is still in the *linear* form. Circularly polarized light is considerably rarer.

4. Direction of polarization: definitions

Let us look at some of the terminology which will be used. Light often originates from some well-defined source but is often subsequently scattered or reflected by particles or surfaces. Such light is visible from parts of the sky other than where the original source is located; and during the scattering or reflection in most cases this light has become more or less polarized.

Tangentially polarized light is light that vibrates perpendicularly to an imaginary line joining the source to the point of observation. Such light can be extinguished by a polarizing filter which would maximally transmit light vibrating parallel to this line. This means that the filter has

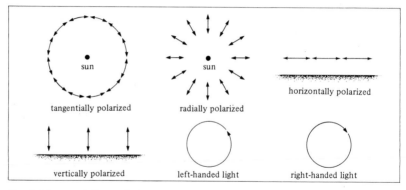

Fig. 9 Tangentially, radially, horizontally, vertically, left-handed and right-handed polarized light.

a radial position with respect to the source (§ 7).

Radially polarized light is light that vibrates *in* the direction of this connecting line. This light is extinguished by a filter in a tangential position.

Horizontally polarized light is light that vibrates parallel to a certain reflecting or refracting flat plane.

Vertically polarized light is light that vibrates perpendicularly to this plane.

These terms refer to *linearly* polarized light. The polarization of reflected light is horizontal and that of refracted light is vertical, while the light of the rainbow is tangentially polarized and the light of the halo is usually radially polarized. In all these examples the vibration direction referred to means that of the (*electrical*) light wave. In addition, as mentioned above, *circularly* polarized light can be left-handed or right-handed. All these possibilities have been illustrated in fig. 9.

Finally, the term *negative polarization* is only used to indicate that the direction of polarization or the sense of rotation is opposite to that which is normally expected on a certain place or object (§ 3).

5. Polarization of light: why does it occur so often?

It is remarkable enough that, merely because of its structure, *unpolarized* light contains all the elements of polarization. Light is a *transverse* wave, which means that the vibrations always occur perpendicularly to the direction of propagation. A light wave *never* vibrates in its direction of motion (this type of longitudinal vibration occurs in sound waves for example). So, when one looks perpendicularly to the direction of motion of a light wave, only vibrations perpendicular to this direction of motion can be seen, *whatever the polarization of the original wave may be* (see fig. 10).

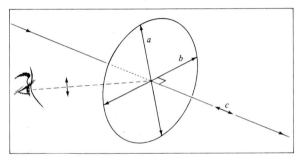

Fig. 10 A wave of light is transverse and so only vibrates perpendicularly to its direction of motion. Vibrations *a* and *b* are possible, vibration *c* is not. Looking perpendicularly at the wave, one only discerns vibrations in one direction (polarization).

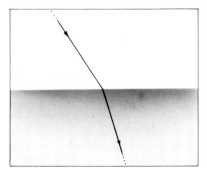
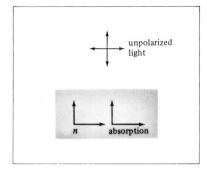

Fig. 11 As light penetrates material, it is refracted: the velocity decreases and the direction changes.

Fig. 12 During transit of unpolarized light through anisotropic material, polarization occurs because the refraction index (*n*) or the absorption is not identical for all directions of vibration.

Thus, when the vibration of light is transmitted perpendicularly to its original direction of motion, linearly polarized light is the result. The transverse character of light waves is itself sufficient, therefore, to convert unpolarized into linearly polarized light. Polarization arising from scattering by particles and from reflection on surfaces is a direct result of this property. And indirectly, the same mechanism also causes refracted light to be polarized. This is because part of the incoming beam is externally reflected as polarized light at the entry face and never penetrates the material, so the remaining part, that does penetrate (and is refracted), must have a reversed polarization.

Conversion of unpolarized light into polarized light is also possible when light penetrates matter and the *material* has a different effect upon light vibrating in various directions. In matter, the velocity of light is less than it would be in air, so that on penetration refraction occurs and the beam of light takes another direction (fig. 11). The degree to which refraction takes place, depends on the index of refraction of the material, n. Also, on its way through the material, part of the light may disappear because of absorption. Now, a great number of materials are *anisotropic*, which means that the index of refraction or the absorption is not equal in all directions of the material. This happens, for example, when the material is stretched or compressed. Consequently, the vertically vibrating part of unpolarized light in fig. 12 is subject to a different index of refraction or absorption coefficient from that affecting the horizontally vibrating component. In the first case, the unpolarized light is split into two polarized beams of light which follow slightly different paths through the material. In the second case, more light disappears gradually from one vibration than from the other so that the beam eventually is converted into polarized light.

Nearly all conversions of natural light into polarized light can be traced back either to the transverse character of light waves (which can be considered as an anisotropy in the vibrations) or to anisotropy of the matter with which light interacts. The same holds for conversions from one kind of polarization into another. In Part III, I shall go further into all these possibilities. On the face of it, it appears that all these conversions are much more likely than, for instance, those from one colour into another: in fact, the chance of a polarized Nature is even greater than of a coloured one.

Observing polarized light

6. With the naked eye

Bees, ants, water-fleas, cray-fishes, fruit-flies, some fish and a number of other animals are able to distinguish between polarized light and unpolarized light as easily as we can distinguish colours. When making use of this ability to orientate themselves, these animals sometimes 'see' the polarization of light at a polarization degree as low as 10%. Man, on the other hand, is almost 'polarization-blind' and generally has to use a polarizing filter to determine the polarization of light. Nonetheless, when light has an extremely high degree of polarization (e.g. > 60%), we can still perceive it with the *naked eye*, although not everybody seems to have this faculty. It turns out that in a plane emitting polarized light, we may see a tiny yellowish figure appear, which we do not see in unpolarized light. It is the so-called *Haidinger's brush* (fig. 13), named after its discoverer. The orientation of this tiny figure depends on the direction of vibration of light

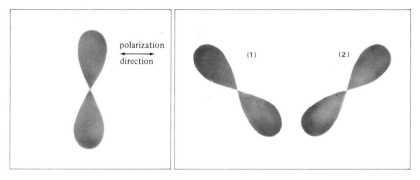

Fig. 13 Haidinger's brush, in the case of linearly polarized light.

Fig. 14 Haidinger's brush in the case of (1) left-handed and (2) right-handed circularly polarized light.

and co-rotates if this plane is rotating. Its longitudinal diameter is about 3°. It appears exactly in the centre of the optical field, hence near the yellow spot of the retina; its origin seems to be connected with the dichroic properties (see §84) of the yellow spot. The Haidinger's brush can be seen, for instance, in the blue sky, which is strongly polarized at 90° from the sun (§15). A similar tiny figure may appear in circularly polarized light; here the orientation depends on the sense of rotation of the light (fig. 14). However, for circularly polarized light, the brush proves a good deal more difficult to observe than in the case of linearly polarized light.

If one wants to visualize the outward appearance of the brush, the best way is to compare it with the Brewster's brush, which one can observe by looking through certain minerals (§70). A photograph of this brush has been given on p. 123 (plate 86). The origins of Haidinger's brush and Brewster's brush are very similar indeed.

One can practise seeing the Haidinger's brush in linearly polarized light by looking through a polarizing filter at a white plane or a white cloud. It will soon be noticed that the centre of the scene is yellowish. On rotating the filter it can be observed that this yellow spot has a structure which rotates together with the filter.

Let us add the following relevant point. As the sun, the moon, the constellations, the rainbows and other phenomena seem larger when they are close to the horizon, so also does the Haidinger's brush. This can be best demonstrated using a linearly polarizing filter: when the brush is observed high in the sky or on a flat surface, a sand plain for instance, it appears distinctly smaller. The explanation of this is the same as for the sun etc. – a spectacular visual illusion!

7. Observing with filters

By far the easiest way to fix polarization of light is by the use of *polarizing filters*. A polarizing filter has the property to transmit either (i) the light of only one direction of vibration or (ii) the light of one sense of rotation. Case (i) applies to a *linearly* polarizing filter with which linearly polarized light can be identified. Such a filter is also called a *polaroid*. Polaroid sunglasses ('anti-glare' sunglasses) have such linearly polarizing filters as 'lenses', which maximally transmit vertically polarized light (fig. 15). Case (ii) relates to a circular filter. When completely polarized light passes through a polarizing filter, the latter is transparent if the direction of vibration of the light or its sense of rotation is equal for both the light and the filter. The filter is opaque to this light and, consequently, appears black when this direction or the sense of rotation is exactly opposite. Thus, a polarizing filter 'translates' polarization in intensity differences for us. So, during the *rotation* of a linear filter the intensity of the transmitted light will vary when

Fig. 15 Linearly polarizing filters are used in anti-glare (polaroid) sunglasses.

the entering light is linearly polarized. Nothing happens, however, when unpolarized or circularly polarized light is going through the filter. Therefore, only linearly polarized light is discernible in this way (plates 4–7).

As previously mentioned, when the polarization of the incident linearly polarized light is complete, the filter is, in one particular position, opaque to this light. In the case of partially polarized light, however, the filter can only reduce the light up to a certain amount. The blurring of a certain linearly polarized object, outlined against a background of unpolarized light, is often striking.

Circular filters are either right-handed or left-handed. Here one has to compare the brightnesses of these two filters to find out whether a certain object emits circularly polarized light. This is, of course, less sensitive than the rotation of a linear filter, so that it is more difficult to observe circularly polarized light. A circular filter generally consists of a linearly polarizing filter with, in front, an additional sheet which converts circularly polarized into linearly polarized light. One should always take care to keep the correct side before the eye. When looked through in the opposite direction, the filter behaves as a linear one. A sheet converting circularly polarized into linearly polarized light is called a *quarter-wave plate* (§93).

By *crossed* filters we mean two filters that are held in such a way that they are *opaque* to each other's light: for example, a right-handed filter and a left-handed filter, or two linear filters one of which has been rotated through 90° with respect to the other (plate 1). *Parallel* filters maximally transmit each other's light. Light that passes through a linear filter is converted from unpolarized light into totally linearly polarized light. A circular filter converts unpolarized light into circularly polarized light. Some filters can polarize light for only a limited number of colours, in which case a change of colour occurs instead of extinction. But 'normal' filters are also not entirely ideal, as weak, blue-coloured light is still transmitted by crossed filters (§84).

The sensitivity of a linearly polarizing filter can be considerably increased by transforming it into a *polariscope*. Minneart's polariscope is constructed as follows: two filters are crossed and a piece of cellophane paper (e.g. from a box of chocolates) inserted between them. In the cellophane zone some light will pass through the filters. Next, the cellophane is turned so that the light transmitted is maximized, and then it is pasted onto the filter. The cellophane rotates the direction of the plane of polarization by a quarter of a turn (fig. 16); it functions as a 'half-wave plate' (§93). When the filter *with the cellophane at the side of the entering polarized light* is rotated, the contrast between the part with and the part without cellophane fluctuates: in a section of the filter the polarizing ability has been turned in the opposite direction. Thus the filter has become a sensitive instrument for observing linearly polarized light (plates 2–3).

The cellophane can also be folded two or three times and put in front of the filter (§53). Then the polarized light shows up in glowing colours. Since

Plate 1 Crossed polarizing filters do not transmit light (§7).

Plates 2–3 Minnaert's polariscope against a polarized background in two positions: a simple but sensitive instrument to fix polarization (§7).

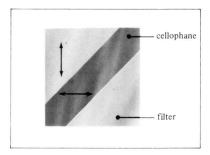

cellophane

filter

Fig. 16 Minnaert's polariscope.

our eyes are more sensitive to colours than to changes of intensity, this method is more refined. The fact that such a filter also reacts to circularly polarized light is, however, a disadvantage. But this is generally not a great drawback, because only a small quantity of circularly polarized light occurs in Nature. On rotating the polariscope, with linearly polarized light

Plates 4–7 The gloss of glass or water is horizontally polarized (§§ 7 and 54); reflection against the metal knife changes it into circularly polarized light (§§ 29 and 48). *Top:* a linear filter can extinguish the glare of the table but the knife hardly shows any difference. *Bottom:* with circular filters the knife shows differences, but the table does not. Arrows indicate the direction of the linearly polarizing filter in front of the camera. L is a left-handed circular filter, R is a right-handed one in front of the camera.

incident, one can observe that the colours change (plates 62–63, p. 94). The *direction* of a linearly polarizing filter can be easily fixed, if it is known that a *reflection* is *horizontally* polarized (plates 4–5).

Besides artificial filters there are also 'natural' filters around us in the form of water surfaces. These reflect vertically polarized light much less effectively than horizontally polarized light. By observing the reflection of an object in water (or in plate glass) its linear polarization can be determined without resort to further aids (§ 15). A polarizing filter can also be made by simply piling up a number of sheets of glass. The pile will prove to function as a linear polarization filter of reasonable quality, providing that one looks through it *obliquely* (§ 79).

While observing polarized light, careful consideration must be given to the following. When looked at obliquely through a plate-glass window at an illuminated object, the light, on its way to the observer, is additionally polarized by its own refractions in the plate glass. Because of this, the observation of weakly polarized light through a window may quickly lead to wrong conclusions, and one should always be aware of this effect. The more obliquely one looks through such a window, the more polarized is the transmitted light. Indeed, the filter of piled sheets of glass described above is based on the very same effect. Car windscreens and aircraft windows likewise affect polarized light; this results in coloured patterns which can frequently be seen through polaroid sunglasses. Here also, we must be careful and preferably look from a short distance through a section of the window which is unlikely to have this property. If possible, we must always try to take our observations outdoors.

Finally, in addition to the above-mentioned usual polarizing filters, there exist more expensive and more sensitive instruments for studying polarized light. In this book, however, we will confine ourselves to those observations of polarized light in Nature which can be achieved with simple sheet filters.

The following terminology is used to describe the filters. Linearly polarizing filters are defined as *polarizers, linear filters* or, alternatively, *polarizing filters* or *filters*. According to the positions of the filters, the following terms are used: *horizontally, vertically, radially* or *tangentially* directed (polarizing) filters (§ 4). Circularly polarizing filters are usually called *circular filters* or, according to their transmission capacity, *left-handed* or *right-handed filters*.

8. Minimum observable degree of polarization

The minimum observable degree of polarization with a polarizing filter depends on our ability to see contrasts in luminosity. This ability is greatest by daylight, and considerably less by moonlight. The situation is most favourable with a polariscope which is rotated before the eye, because the contrast of luminosity changes directly and simultaneously through and beside the cellophane. Using a normal linearly polarizing filter, we see the brightness vary when the filter is rotated backwards and forwards through a quarter-turn, but we have to remember the observed intensities for a short time in order to fix the polarization. Hence this method is less sensitive than using a polariscope. To observe circularly polarized light we must even compare the differences in contrast between two filters which are never entirely identical. This is the most insensitive technique. Hence circularly polarized light is always less easily observable than linearly polarized light. We can adhere approximately to the following standard:

Method	Minimal P (or q)	Method	Minimal P (or q)
Haidinger's brush	$> 50\%$	Circular filter	20%
Linear filter	$10–15\%$	Linear filter by moonlight	50%
Polariscope	$5–10\%$	Polariscope by moonlight	30%

The perceptibility of polarized light also depends upon whether its *background* is polarized or not. When the direction of polarization of light from the background is different (e.g. a $46°$ halo in the blue sky), the situation is much more favourable than when they are in the same direction (e.g. a cloud against the blue sky), because in the latter case light from both object and background is reduced with a filter. The use of a polariscope is preferable in such a case, because with an ordinary filter one often looks unintentionally at the difference of contrast between the object and its background.

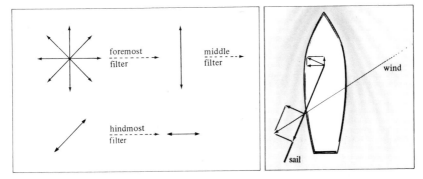

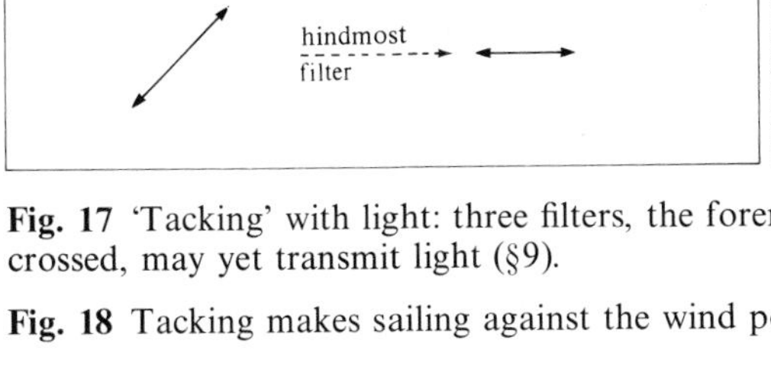

Fig. 17 'Tacking' with light: three filters, the foremost and the hindmost crossed, may yet transmit light (§9).

Fig. 18 Tacking makes sailing against the wind possible.

9. 'Tacking' with light: a test with three polarizing filters

In this paragraph, I shall describe a simple test with linearly polarizing filters, which will clearly show that there is 'something special' about polarized light.

First, two linear filters are crossed so that they do not transmit any light. Then a third filter is placed so that its transmission direction is at 45° to the first. If this third filter is inserted *between* the other two filters, the sequence transmits light again!

The cause is the so-called double (vector) decomposition of the polarized light by both the middle and last filters – the same principle which allows sailing boats to be navigated obliquely against the wind. The middle filter is not crossed with respect to the first and so the combination transmits light. The light transmitted vibrates, however, at an angle of 45° with respect to the hindmost filter so that it is also partially transmitted by this filter. This would not occur without the middle filter. This phenomenon is illustrated in fig. 17. The same occurs in the case of a boat that is sailing in the teeth of the wind (fig. 18): the force exerted by the wind on the sail is first decomposed into a force parallel to the sail (this force no longer has any influence) and a force perpendicular to it. Since the sail is slanting, the latter force can again be decomposed into a force perpendicular to the boat and one parallel to it. Again the former does not play a role but the remaining force causes the boat to sail. However, as this force now has a component opposite to the direction of the wind the same holds for the direction of sailing. When the sail is parallel to the boat, this double decomposition does not occur and it is impossible to tack against the wind. In view of this parallel between sailing and the transmission of light through three filters, we can reasonably speak of a 'tacking' manoeuvre to transmit polarized light.

10. The history of the discovery of polarization

This part concludes with a chronological survey of the history of the discovery of polarization till 1973. It focusses mainly on the polarization of light in Nature, as discussed in this book, but other facts are also mentioned. The survey consists partly of data taken from Gehrels, partly of information from other sources.

About 1000 The Vikings discovered the dichroic properties of crystals like cordierite. With these crystals they observed the polarization of the blue sky and were thus able to navigate in the absence of the sun.

1669 Erasmus Bartolinus from Denmark discovered the double refraction of calcite crystals.

1690 Huygens discovered the polarization of the doubly-refracted rays of calcite, without, however, being able to explain the phenomenon.

1808 Malus found the polarization of reflected light by using a calcite crystal as a filter. This filter apparently loses its double refraction when the entering light is polarized and the crystal is held in the correct position. Afterwards Malus formulated his law, which gives the relationship between the position of a polarizing filter and the quantity of transmitted light, when the entering light is totally (linearly) polarized.

1809 Arago rediscovered the polarization of the blue sky and later found the neutral point, which was named after him. In 1811 he discovered the optical activity of quartz, and in 1812 he constructed a filter out of a pile of glass sheets. In 1819 he found the polarization of comet tails and in 1825 the (weak) overall polarization of 22° haloes. In 1824 he found the polarization of the glow emitted by hot, incandescent metals. He was also the first to record the polarization of the moon.

1811 Biot discovered the polarization of the rainbow. In 1815 he established the optical activity of fluids such as turpentine, and in 1818 he studied the optical activity of gaseous turpentine in a gas column of 15 m length. Unfortunately, this apparatus exploded before he could finish his meesurements. In 1815 Biot also discovered the strong dichroism of tourmaline.

1812 Brewster discovered the law, which was named after him, that indicates the relationship between the index of refraction and the angle of incidence at which light is totally converted by reflection into linearly polarized light. In 1818 he discovered Brewster's brush in pleochroic crystals, and in 1842 the neutral point in the blue sky which bears his name.

1816 Fresnel gave a theoretical explanation of the existence of polarization.

1828 Nicol invented his prism, which can be considered to be the first easily usable polarizing filter.

1840 Babinet discovered the neutral point in the blue sky named after him.

1844 Haidinger found that the human eye has the ability to distinguish between unpolarized and polarized light, because in the latter case a yellowish figure appears on the retina (the Haidinger's brush).

1845 Faraday discovered the rotation of the polarization plane in magnetic fields.

1852 Herapath made a synthetic crystal with very high dichroism, which was the first step on the way to the construction of simple sheet polarizers.

1858 Liais discovered the polarization of the solar corona.

1860 Kirchhoff found that incandescent tourmaline emits polarized light, according to the radiation law which he also formulated.

1874 Wright discovered the polarization of zodiacal light.

1884 Kiessling recorded that the glory is polarized.

1889 Cornu found that artificial haloes in sodium nitrate crystals are highly polarized because of the double refraction of the crystals.

1905 Umov described the relationship between the degree of polarization of light reflected from rough surfaces and the albedo of the surface.

1911 Michelson discovered that certain beetles have a gloss which is circularly polarized.

1928 Land constructed his first polarizing filter. Further developments of this filter made it possible to study effects of polarization with a simple and efficient sheet filter. Such filters are also used in sunglasses etc. to reduce the intensity of glare. Compared with the Nicol and other crystal filters used up to that point the development of this kind of sheet filter meant great progress.

1935 Beth posited that circularly polarized light exerts a slight mechanical torque on materials and thus proved directly the rotating character of this light.

1939 Le Grand and Kalle reported that scattered light underwater is polarized.

1940 Bricard found that supernumerary fog-bows shift when one looks at them through a linear filter which is then rotated.

1947 Van de Hulst gave the first feasible explanation of the glory and explained its polarization directions.

1949 Von Frisch discovered that bees are more capable than man of distinguishing polarized from unpolarized light and use this ability to orientate themselves.

1949 Hall and Hiltner found that the light of stars is slightly polarized.

1954 Dombrovsky discovered the strong polarization of the Crab Nebula.

1955 Shurcliff discovered that the human eye is also capable of distinguishing circularly polarized from unpolarized light.

1956 Jaffe proved that, when the egg cells of certain algae are irradiated by linearly polarized light, they tend to develop in the direction of vibration of the light.

1958 Duncan discovered the polarization of the aurora.

1960 Witt established unambiguously the polarization of noctilucent clouds.

1973 Shaw measured the polarization distribution of the sky during a total solar eclipse and found a symmetry with respect to the zenith.

1973 Gehrels edited the first general source book on polarimetry.

Plate 8 Since the first satellite was launched in 1957 the phenomena of light and polarization have been studied from space. This photograph was made in 1975 by the American weather satellite Essa 8 from an altitude of 1420 km. Above to the left, the snow-covered continent of Greenland is visible. To the right of it, in the centre of the frontal cloud-cover over the ocean the sub-sun appears (§ 39) as a bright spot, indicated by arrows.

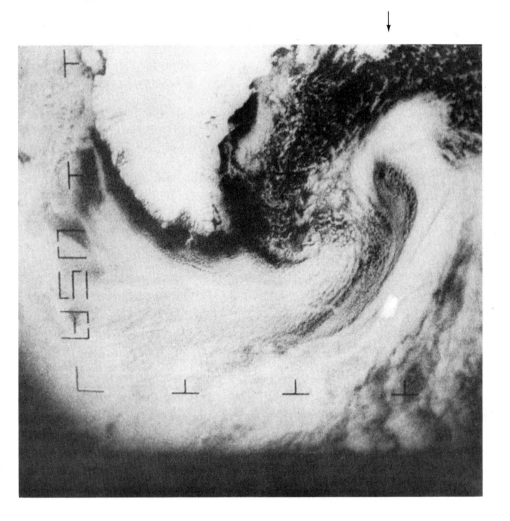

Part II

Polarized-light phenomena in Nature

Light and illumination, colour and polarization in the natural environment: a survey

11. Light, colour and polarization

Although unpolarized sunlight is by far the most important source of the light which we observe around us, yet it is not this light itself that determines the colour and the polarization of the scenery, objects and sky. The way in which this manifests itself depends on how objects and particles transmit sunlight to us: the light intensity has been considerably reduced, the light often has colour and as a rule it also proves to be more or less polarized. The manner and the measure in which this polarization occurs, depends strongly on the path the light has taken: it could have been transmitted via one object directly to our eye (e.g. by reflection) but it could also have followed its path to the observer via many particles or objects, each of which modify the polarization of the light.

A comparison between colour phenomena and polarization phenomena is unavoidable, for, while white (uncoloured) sunlight results in colours in Nature, the unpolarized sunlight similarly gives rise to much polarized light around us. The quantity of polarized light is at least as large as the quantity of colour. However, conversions of sunlight into coloured and polarized light are usually not parallel so that white light can be polarized but coloured light need not display polarization at all. Consequently, shades of polarization are quite different from the colour shades in Nature. Unlike differences in colour, the differences in polarization in the scenery are hardly noticed by Nature-lovers; in fact, they are barely visible to the naked eye. With a polarizing filter, however, we are suddenly aware of the many nuances in polarization phenomena.

The above-mentioned comparison between colour and polarization will only serve as long as the polarization is linear. *Circularly* polarized light is fairly rare in Nature; only very seldom is it formed directly from unpolarized light. Commonly it originates indirectly in a *modification*

25

sequence by which sunlight is first converted into linearly polarized light and then into circularly polarized light. However, this only happens if by coincidence all these successive modifications co-operate in the right way. The fact that nevertheless circularly polarized light sometimes does occur conspicuously in Nature, adds a subtle dimension to the diversity of light phenomena in the open field.

Not only the *nature* of the objects that transmit light to us but also the way in which the scenery is *lit* influences the polarization of the light around us. In principle, however, the situation is the same for illumination by the sun or the moon: in both cases there is one predominant source of light shining upon the scenery and the air from one well-defined direction. By fainter moonlight the eye is a good deal less sensitive to the differences in brightness which appear through a polarizing filter. So our environment *seems* to contain less polarized light. Similarly, the moonlit scenery also seems less colourful: it is only an illusion caused by the considerably reduced colour sensivity of the eye under these circumstances. For instance, by moonlight the sky is as blue as by sunlight. We do not see it, however; to us it seems a whitish haze before the stars.

On the other hand, an overcast sky is really another form of illumination. It is a diffuse light source and hence shadows are barely visible. Under these circumstances Nature is definitely less polarized than under cloudless skies. On closer inspection, in this situation more light proves to come from the zenith (the point exactly above the observer) than from the horizon; this leads to some remarkable effects of polarization. The situation is indeed worth being studied, as the directions of polarization and the degrees of polarization differ so much from those under cloudless skies. Another possibility is the illumination of the scenery during twilight in the presence of only few clouds. Here, the source is also diffuse but now most of the light comes from the horizon. We can observe striking effects of polarization, because the blue sky itself, being responsible for this illumination, is already strongly polarized. In this situation the quantity of polarized light is considerable. There is also more circularly polarized light than in bright sunlight, because in the above-mentioned modification sequence (unpolarized light → linearly polarized light → circularly polarized light) the first step is no longer necessary. Finally, the illumination during a total solar eclipse is a form of ghostly twilight, in which changing and spectacular polarization phenomena can occur.

There are of course only faint boundaries between the three possibilities of illumination (sunlight, clouds and twilight). A flat surface (e.g. of water) will often reflect a part of the blue sky in the observer's direction instead of the sunlight itself. In this case the manner of illumination is comparable to the situation at twilight, and analogous effects can occur. In most cases, however, the tripartite division holds.

By far the greater part of the light in our surroundings arises from reflection or scattering. In the case of direct illumination, e.g. by the sun or the moon, it usually leads to *tangentially* polarized or *horizontally* polarized light (when a reflecting surface is very smooth). The polarization is generally maximal for objects in a position of about 90° from the sun, though there are exceptions to this rule. But the blue sky, for example, has its maximum polarization at 90° from the sun. Under *overcast skies* the polarization of light is mostly *horizontal*; at *twilight* it depends greatly upon the direction of observation and whether the reflected light originates from strongly or less strongly polarized parts of the sky. In all these cases polarization is not, or is hardly, dependent on colour, so that we can certainly extinguish the light with a polarizing filter, but on doing so we do not usually see any change in colour.

In addition to scattering and reflection there are a number of processes that can cause polarization of natural light, like refraction, double-refraction, surface waves, etc. All these processes produce only a relatively small part of the total quantity of polarized light in Nature, but owing to the difference in form of manifestation, in the direction of polarization or because of the occurrence of circularly polarized light, they are a separate group among the polarization phenomena around us. Moreover, light that has already been polarized can subsequently be modified by processes like total reflection, reflection on metals and in many other ways. This usually also results in altered directions of polarization and sometimes in formation of circularly polarized light, and consequently leads to a greater variety in the polarization of the light in our surroundings.

Colour-dependent polarization effects arise in a magnificent way when polarized light is modified by doubly-refracting or optically active materials. Here our polarization-blindness is manifested in its most spectacular form: with the help of a polarization filter transparent matter suddenly displays the deepest graduations in colour. Many minerals (e.g. ice-crystals) show this effect; the same holds for car windscreens and stretched materials. The colours thus visualized are among the most splendid Nature can offer us.

The foregoing discussion gives some idea of the enormous variety of polarization phenomena that can occur in Nature. It is remarkable when we realize that all these shades remain almost invisible to the naked eye, but can be revealed in all their beauty with the aid of a simple polarizing filter. Subsequent transformations of polarized light can increase the already large number of possibilities; moreover, contrasts in polarization (i.e. differences in polarization between objects and their background) can sometimes to a surprising degree affect the appearance of phenomena viewed through polarizing filters. It appears that some polarized sources of light exist which enable us even at night to take observations of a special class of objects. All these possibilities will be described in detail, as if we were going for a walk in the open field, armed with polarizing filters in order to reveal to ourselves this facet of light phenomena in Nature.

The blue sky and the clouds

12. The blue sky

The sky is blue when it is cloudless – a marvellous natural phenomenon. If the blue sky were as rare as a rainbow, we would be much more aware of its beauty – as it is, we usually appreciate it only after days of overcast weather.

The blue light is sunlight scattered by the molecules of the atmosphere. Thus outside our atmosphere the sky is inky and the stars are always clearly visible, as astronauts know from experience. The atmosphere is rather thin (some tens of kilometres) and looking straight upwards, our eyes will meet fewer constituent molecules of air than when we look at the horizon. That is why less light comes from the zenith than from the horizon; moreover, the sky straight above us is deeper blue, the more so as we are at greater height because the layer above us is less thick. Near the horizon the sky is lighter blue and it often inclines to a whitish tint. A sharp observer will also find that the sky at 90° from the sun is less luminous than anywhere else. This has nothing to do with the thickness of the scattering layer, but it results from the effects of polarization and the scattering behaviour of the molecules in the air.

The blue sky is strongly tangentially polarized (plates 9–15). The strongest polarization is at 90° from the sun, and at twilight its degree of polarization can reach a maximum value of about 75% (§73). When we look closer to the sun, the degree of polarization is less: at 45° from the sun it is 30% or less, and very close to the sun it is 0%. A similar decrease in the degree of polarization is seen when we look gradually at the area of the sky straight opposite to the sun. It appears that the strongest polarization occurs when the air is very clear; a hazy sky is a great deal less polarized. At sunrise or at sunset the polarization along almost the entire horizon is directed vertically (see also §14), since tangential corresponds to vertical in this case. At a high solar elevation, the direction of polarization with

29

respect to the horizon depends greatly on the point of the compass at which we are looking: opposite to the sun, for example, the polarization is horizontal. In this situation, the degree of polarization is still very high, and it can amount to 50–60 %. Watching the sky with a polarizing filter which is directed in such a way that the light has been maximally extinguished (i.e. radial to the sun), we see an impressive dark band appearing at 90° from the sun, which marks the area of the maximum polarization (plate 9). The width of the band is about 30°; its colour is deep blue (§§ 13, 86). In the band the sky can be up to eight times brighter in the tangential direction than in the radial direction. Some *insects* (bees, ants etc.) do not make use of the position of the sun but use the polarization distribution of the sky to find their way: they are able to see the polarization of light many times better than we do (§6), although this capacity seems to be mainly restricted to ultraviolet light. In the absence of sunlight, the *Vikings* also used this method of navigation: they inferred from the polarizing direction of a small piece of the blue sky where the sun must be, and from this they plotted their course. They may have applied this method even in fog or when overcast by thin clouds, since the sky retains its polarization pattern under these conditions also (see §24). They probably used a cordierite crystal as a polarizing filter, which colours blue instead of light-yellow when it points to the sun (as a consequence of the polarization of light from the sky). They called such crystals 'sunstones'.

With a polarizing filter in tangential position the sky appears to be more or less uniformly bright when we are watching at any particular elevation from the horizon; with a radially directed filter we can see the image as outlined above. Without a filter we see the sum total of the two directions of polarization, which results in a somewhat lower intensity of light at 90° from the sun. This minimum of light at 90° from the sun which was

Plates 9–10 *Left:* with a polarizing filter in the correct position one can see the strongest polarized area of the blue sky silhouetted against the sky as a dark band (§ 12). *Right:* reflection of this light results in a dark spot on the water, visible without a filter (§§ 15, 48 and 54). Both photographs are taken with a wide-angle lens.

Plates 11–13 Hazy sky at sunset. *Above:* without filter. *Below:* with polarizing filter in two positions. With the filter the colour of the sky can be switched from white to blue, especially at 90° from the sun (§13). At the same time there is a change of contrast between sky and clouds (§16), while the colour of the clouds becomes more yellow (§20).

mentioned earlier is therefore a result of the strong polarization of the light from the sky.

13. Haziness on the horizon: colour effects with a polarizing filter

In the case of slight haziness in the atmosphere, the colour of the sky near the horizon is white instead of blue. Now too the light is strongly linearly polarized, particularly at 90° from the sun. We can therefore extinguish this light with a polarizing filter, but on doing so the colour *switches* from whitish to deep blue (plates 12–13). This is particularly striking when the sun is at a low elevation.

The reason for this is as follows: like the gas molecules in the atmosphere, the minuscule particles causing haziness scatter mainly *bluish light*, but they are much more effective scatterers. On its long track toward our eyes the scattering light has a good chance to be scattered again by other particles. During all these scatterings the light is gradually losing its blue colour, because exactly this light is most likely to be 'intercepted'. So the primary scattered light is turning whiter and whiter, but its strong polarization remains. On the other hand, light that has been scattered out of the beam may still reach us via additional scatterings, but it will then have largely lost its polarization (§86). However, once again because blue light is most likely to be multiply scattered, the light has a deep-blue colour. Extinguishing the white singly scattered light with a polarizing filter therefore causes the slightly polarized multiply scattered light to be immediately clearly visible and the sky turns blue again! The same mechanism can also be applied to the clear blue sky, which becomes a very deep blue colour if a polarizing filter is used to extinguish its light maximally (§12).

Light scattered from thicker haze also remains polarized. We can make use of this, with the aid of a polarizing filter, to improve the visibility of distant objects at about 90° from the sun in haze (see also §§16 and 47).

14. Twilight and negative polarization

The blue sky reaches its maximum polarization when the sun is just below the horizon. In this situation and also when the sun is low, it is remarkable to find that the entire illumination is caused by the blue polarized sky. Its polarization is, as mentioned, maximal at 90° from the sun and it decreases towards or away from the sun. The highest possible degree of polarization occurs in the zenith. Over the entire firmament the average degree of polarization is about 40%. If all light from the sky were concentrated in one point, this would consequently yield its degree of polarization.

Let us make the following observation at twilight: trace the polarization of the sky by starting at the zenith and stopping at the point of the horizon that is exactly opposite to the sun. It will be found that the polarization diminishes and the degree of polarization is zero at a point that is about 25° above the horizon. Closer to the horizon the degree of polarization is again increasing, but now the direction has become *reversed* and it is therefore oriented perpendicularly to the horizon! This switch in the direction of polarization in the sky is called *negative polarization* (§4); at twilight it reaches a maximum degree of polarization of about 20–30%. The point at

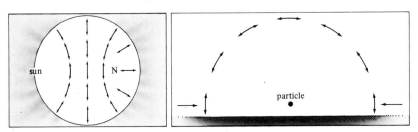

Fig. 19 Directions of polarization in the firmament at twilight. N is the neutral Arago point.

Fig. 20 The way an air particle would see the polarization of incident light at twilight. Most light comes from the horizon.

which the polarization degree is zero, is called the *neutral Arago point*. The result of this negative polarization is that almost everywhere along the horizon the direction of polarization is vertical, including places where one would have expected it to be otherwise (see fig. 19). Isolated clouds, in the region of the sky opposite to the sun, show this altered direction of polarization as well (see §23). The unexpected switch in the direction of polarization, if one looks from zenith to horizon, is a gratifying object of observation for the amateur; but the effect is best observed with a polariscope.

The phenomenon is explained by the fact that in this area the firmament is not lit by the sun but by the blue sky itself. A molecule or particle 'sees' therefore merely light from the sky and not from the sun. Light from the zenith is horizontally polarized and that from the horizon vertically polarized (fig. 20). If the same quantity of light were to come to the particle from all directions, the sum of all this light would be unpolarized and would be transmitted as such by the particle. However, the greater part of the light comes from the horizon and thus has a preferential direction. So the scattered light is perpendicularly polarized to this average source of light, hence vertically polarized. As the entering light is already, on average, vertically polarized (the vertically polarized light from the horizon predominates over the horizontally polarized light from the zenith), the vertical polarization of the secondary scattered light is more pronounced. Clearly, negative polarization should not occur in single scattering, but is once again a result of multiple scattering (§86).

It appears that such an area of negative polarization also exists in the direction of the sun. But here the polarization degree is much less ($\sim 6\%$), because after sunset the greater part of the light still results from single scattering. The attendant neutral points are called the *Babinet point* (above the sun) and the *Brewster point* (below the sun). These types of points have been the subject of extensive studies – the latter point is the most difficult to observe. When the sun is rising above the horizon, both areas of negative polarization remain but the degree of polarization of the former drops so rapidly to 6% or lower that it also rapidly becomes impossible to observe. To us, the spectacular switching of the direction of polarization in the celestial area opposite to the sun at twilight is the most interesting phenomenon arising from negative polarization.

15. Observation *without* filters of the polarization of light from the sky

It is possible for an acute observer to ascertain the strong polarization of the light from the sky without any artificial aids but with the naked eye alone. Usually, it is best carried out when the position of the sun is low. There are four ways of observing the polarization:

1 Haidinger's brush. The polarization of the sky is sufficiently strong for this observation. The best way is to look straight upwards at twilight and to move slowly to and fro, in such a way that one's chin points by turns westwards and southwards. The tiny yellowish figure can be seen which betrays the presence of polarized light to the naked eye (see §6); this figure points like a compass needle towards the sun. If one remains motionless, the brush will fade away. But because the head is moving, this does not happen as the figure keeps appearing on different parts of the retina.

2 Reflection of the blue sky in a smooth surface (e.g. water) (§7). At 90° from the sun and at low solar elevation, light from the sky near the horizon is vertically polarized but still water and other flat surfaces best reflect horizontally polarized light (fig. 21). Thus an 'unnatural' dark spot can be seen in the water at 90° from the sun indicating the polarization of the light from the sky (plate 10). This observation is possible with still water, shiny car roofs and wet or dry asphalt roads; one must always look downwards at rather a steep angle in order to see the spot. A water surface is clearly more transparent at this spot because there is hardly any gloss (see also §§48 and 54). In ditches with slightly rippling water this dark spot is also distinctly visible, particularly when one is passing them rapidly, by train for instance. Then the spot makes the impression of being more or less triangular.

3 Reflection on car windscreens. These windscreens are doubly-refracting and coloured patterns appear on them, caused by chromatic polarization (§93). One does not need polarizing filters to see the patterns, because the light is internally reflected by the window and the blue sky acts as a source of polarized light (fig. 22; see also §52). When the sky is overcast the patterns can therefore hardly be seen. Plastics

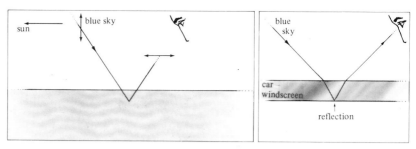

Fig. 21 At twilight the vertically polarized light from the sky is only poorly reflected by a (horizontal) water surface; therefore at 90° from the sun the water surface is dark.

Fig. 22 Because of the internal reflection of the polarized light from the sky in the glass, coloured patterns can be seen in car windscreens without the aid of a polarizing filter; thus the polarization of the blue sky is betrayed to the naked eye.

Plates 14–15 The contrast between the blue sky and the clouds can often be entirely reversed with a polarizing filter (§ 16). It can best be seen when the sun is rather low, at about 90° from the sun.

and other materials may also show this effect. It also occurs when the sun is high; usually the patterns can be a good deal intensified by using a polarizing filter.

4 Finally, the observation (without a polarizing filter) that the sky at 90° from the sun emits less light (see § 12) provides an indirect means of finding the polarization of the light from the sky with the naked eye.

16. Half-clouded sky and objects in the air: polarization contrasts

The beauty of polarization of the light from the sky is at its height when cloudbanks, e.g. altocumulus, are outlined against the bright blue sky. These clouds are also tangentially polarized with respect to the sun (see § 20), but their degree of polarization is much less than that of the sky. The consequence is that on rotating the polarizing filter we can observe an enormous change in contrast between the clouds and their blue background. This is most spectacular when the clouds are slightly darker than the sky and stand out greyish against it. The contrast can be even *switched* with a polarizing filter in this case because then the light from the sky in a radial direction is fainter than the light of the clouds. We see the cloud outlined against the deep-blue sky as *white* instead of *greyish*! Without doubt these polarization contrasts belong to the most surprising phenomena which can be observed with a polarizing filter (see plates 11–13 and 14–15).

For water-clouds, which consist of small drops of water, there is another remarkable phenomenon. For the clouds are rather strongly polarized not only at 90° but also at about 145° from the sun (see § 21). This polarization is, however, stronger than that of the background sky; which results in the polarization contrast being the *reverse* of the situation at 90° from the sun. With the polarizing filter in a radial direction these clouds seem to be white at 90° and grey at 145°; in a tangential direction it is quite the reverse. This opposing behaviour is obviously not the consequence of opposite directions of polarization in these two areas of the sky but of an opposite *contrast* of polarization between the clouds and their background. With regard to ice-clouds, there is no strong polarization at 145° and therefore the switch of contrast does not occur.

Maximum extinction of the blue sky with a polarizing filter often makes objects in the air become more clearly defined. This applies to (thin) clouds, the moon, smoke, birds, remote aeroplanes and mountains, kites, drift-sand etc. (see also §§ 13 and 47). Other objects, such as dark-grey clouds, which in this situation may become about as bright as the background sky, are clearer with the polarizing filter in the other direction, as they are then silhouetted against a bright sky.

17. Visibility of stars and planets by day

Besides the sun and the moon, Venus, Jupiter and Sirius are the brightest celestial bodies. When the planet is brightest Venus can be seen easily with the naked eye in broad daylight. Its position, however, must be known precisely, so it is better to look for it first with a pair of simple binoculars. This type of observation is most easily done when the moon is close to the planet and thus can act as an orientation point. It is much more difficult to see Jupiter by daylight; nevertheless, it is possible to observe it with the naked eye when the sun is low, and also through binoculars when the sun is high but only when the air is very clear. Finally, Sirius is a borderline case but it is just visible when the sun is low.

One would expect that it would be easier to observe stars and planets with a polarizing filter by day, because particularly at 90° from the sun the light from the sky can be extinguished to a large degree and the celestial bodies themselves are unpolarized (§58). In practice, however, this method proves to be disappointing: weakening of starlight with a filter makes observations disproportionately more difficult. One learns by experience that immediately after sunrise Jupiter can hardly be seen any better with a polarization filter than without, even at 90° from the sun. The situation is even more unfavourable at twilight, and the visibility is decreased rather than increased with a polarizing filter. As Venus can never appear further than 46° from the sun and is therefore never in the most polarized area of the sky, an attempt to improve its visibility by using a polarizing filter does not make any sense at all.

It can not be excluded, however, that there might be situations in which a polarizing filter is a help for this kind of observation. Probably the best circumstances are when the solar elevation is not too high. One should use a polarizing filter which is as transparent as possible; any extra loss of light is detrimental. If the filter is helping, in theory Jupiter and perhaps Sirius should become just visible. For Sirius it is best to try in October, when it is at 90° west of the sun and can be traced at dawn.

18. The moonlit sky

A moonlit sky is as blue and polarized in the same way and to the same degree as the sunlit sky. By dim moonlight, however, the eye is virtually insensitive to colours and also less sensitive to contrasts in brightness. That is why we now see the 'blue' sky merely as a whitish haze hanging before the stars, and its polarization too seems less. Yet with a polarizing filter or a polariscope we can ascertain that the polarization of the moonlit sky is substantial. The observation is easier at full moon than at the first quarter, because in the latter case the moonlight is fainter by a factor of nine and so is the resulting light from the sky. It is easy to ascertain with a polarizing filter that the characteristics of the polarization are identical with that of the sunlit sky and, for example, is maximal at 90° from the moon.

The polarization of the moonlit sky can also be observed from the visibility of the stars (which are unpolarized). It is easy to establish that a polarizing filter which maximally extinguishes the light from the sky, shows the stars better than when it is in the reversed direction; hence more stars are observed. However, still more stars can be seen with the naked eye; under these circumstances the use of a filter does not improve their visibility at all. From all this, it is evident that viewing the stars through filters or sunglasses makes their visibility deteriorate much more than might at first be expected (see also § 17).

19. Total solar eclipse

This is one of the most imposing natural phenomena; unfortunately, it is also very rare. One usually has to travel far in order to observe it. During such an eclipse, the sun is *completely* covered by the moon for at most some minutes, and this immediately transforms the aspect of the sky completely. When the last part of the disc of the sun disappears behind the moon, it seems as if the 'light' has been 'switched off': in one second it is as dark as if it were just an hour after sunset (when the sun is 5–7° below the horizon). At the same time a breeze rises: the eclipse wind. Instead of the white solar disc an inky disc surrounded by a silvery white aureole is seen in the sky; the aureole is called the *solar corona*. It is about as bright as the full moon. The light of the corona is scattered sunlight; the scattering takes place in space on free electrons which surround the sun like a cloud. This scattered light is highly polarized (§ 73); the maximum degree of polarization is about 40% at a distance of a quarter of the solar diameter from the edge of the sun. Farther from the sun the degree of polarization is gradually decreasing. This polarization is tangentially directed. Looking straight into the corona with a polarizing filter in different positions, we find that its shape is clearly *changing* (plates 16–17). Taking the corona as a whole, the directions of polarization more or less neutralize each other, with the result that its *total* radiance shows hardly any polarization. The sky itself, however, is not lit up by the radiance of the corona alone, for then it would have been a thousand times darker. Its main source is light coming from *outside* the area where the totality is taking place and where the sun is still shining in abundance: this is never farther away from the observer than 130 km, because these zones are so small. In fact, a particular type of twilight now forms. Most light is seen near the horizon where parts of the atmosphere are still lit by the partially eclipsed sun outside the zone of totality; the sky is darkest in the zenith. It is often so dark that the brightest stars are visible.

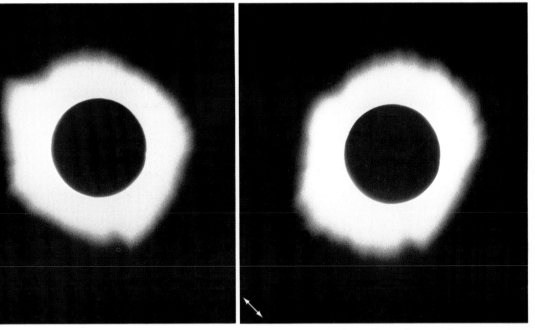

Plates 16–17 The solar corona, visible during total solar eclipses, is strongly tangentially polarized. So its shape alters with a polarizing filter: it becomes more or less stretched in one direction (§19). (Photographs Astronomical Observatory, Utrecht, 9 March 1970.)

As the brightness and the distribution of light of the sky are immediately transformed at totality, so also is its polarization. The polarization is now no longer tangential with respect to the sun, because the latter does no longer act as the main source of light. As outlined above, the light now comes from the horizon and originates from regions where the sun is not totally eclipsed (fig. 23). Just as in the case of ordinary twilight, the source of light can be thought to be on the horizon and an additional scattering by air particles results in vertical polarization (§86). But since this multiply scattered light is now completely dominant, its vertical polarization shows

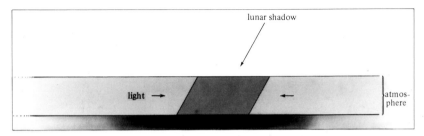

Fig. 23 During a total solar eclipse the sky is lit up by scattered light coming from outside the zone of totality.

up all over the sky; in the zenith the polarization is absent or very low. The highest degree of polarization of the sky is at about 30° above the horizon and may be 30–40%. The distribution of the degree of polarization is symmetrical with respect to the zenith.

After a few minutes the totality ends and the illumination is normal again; a very short time indeed for taking in all the accompanying phenomena. It may be many centuries before a total eclipse of the sun will again be discernible in the same place.

20. Polarization of clouds: general features

The light of clouds is linearly polarized and the polarization depends upon the illumination. When the cloud particles are directly lit by the sun or the moon (i.e. partly clouded sky, thin clouds or fog), the direction of polarization is tangential, as in the case of blue sky. But the degree of polarization is usually a good deal less. We can observe altered directions of polarization appearing by twilight or under an overcast sky.

There are a great many kinds of clouds but the polarization of the light of the clouds depends only on their illumination, their composition (ice-crystals or drops of water), their density and on the size of their elements.

Although the direction of polarization of clouds is the same as that of the blue sky, its degree of polarization may differ considerably, so that through a polarizing filter we can see great contrasts between the polarization of clouds and that of the sky at 90° from the sun (§ 16). The tangential polarization of the white light of clouds is itself somewhat disguised by this, and this can only be observed with close scrutiny. Problems can be avoided by the use of a polariscope, with which it is easy to see the tangential polarization of the clouds. It is interesting that, with a filter and especially at 90° from the sun, it is possible to see a *colour change* in the clouds: when their light has been maximally extinguished, the clouds will turn slightly yellower (see plates 12–13). This colour change can best be seen when the sun is not too high and the atmosphere is somewhat turbid. The effect is caused by the column of air between the cloud and the observer: the column also emits bluish scattered light, which slightly changes the hue of distant objects (see also §47). Extinction of this tangentially polarized scattered light causes, therefore, a yellowing of the background (of the cloud in this case).

21. Ice-clouds and water-clouds

Roughly speaking, these are polarized in the same way – tangentially with the highest degree of polarization about 90° from the sun, exactly like the blue sky. The maximum degree of polarization does not, however, amount to more than approximately 40%; the polarization decreases when one is looking closer to or further away from the sun.

There is one striking difference between the polarization of sunlit water-clouds and that of sunlit ice-clouds. At about 145° from the sun the former show a sharp rise in polarization which does not occur in the case of ice-clouds. Here, the degree of polarization can easily amount to 60% and is, therefore, even higher than its value at 90° from the sun; with a rotating polarizing filter the clouds can be clearly seen becoming brighter or extinguished in this part of the sky. Even without filters we can see that something strange is happening here: the clouds are markedly brighter. This can be observed best when the sunlit cloudcover is more or less uniform, e.g. altocumulus banks (see also §30 and plate 30 on p. 55).

This phenomenon is caused by *rainbow-scattering*: a rainbow can just as well be formed in minuscule cloud drops as in rain drops, but here it is completely white and therefore not conspicuous. It does show, however, the same polarization as occurs in the rainbow itself (see further §30). Of course the area of the sky where it occurs is only visible at a rather low position of the sun, because otherwise it is below the horizon. This high polarization leads to altered contrasts of polarization between the sky and the clouds, as described in §16.

As the degree of polarization of clouds depends on so many factors (§20), it may differ from cloud to cloud; for instance, polarization is often found to be less if the clouds are denser. However, the above brief description of the polarization of clouds approximates well with what can often be seen through a polarizing filter.

22. Polarization contrast between ice-clouds and water-clouds. Dust-clouds, sand-clouds and smoke

The magnificence of the difference in degree of polarization between ice-clouds and water-clouds at 145° from the sun can be seen when a low sunlit water-cloud is in front of an ice-cloud. Cirrus is, for instance, an ice-cloud, but the vivid white cumulonimbus clouds may also consist of sufficient ice-particles that their polarization behaves in the same way. Indeed, from the point of view of polarization, this cloud can often be taken to be an ice-cloud. Through a tangentially directed polarizing filter we can see that the water-cloud and the ice-cloud behind it are more or less of the same brightness, but in a radial direction it is only the water-cloud that is considerably extinguished. It stands out greyish against the striking white cumulonimbus. Along with the polarization contrasts between clouds and sky (§16), this form of polarization contrast belongs to the clearest that can be observed in Nature. However, one has to be sure to look in the right direction – at 145° from the sun. It is not visible anywhere else.

Dust-clouds and *sand-clouds* have a polarization direction which is generally the same as that of ice-clouds, only the degree of polarization is usually less. So here their polarization contrasts with the clouds in the sky behind them: these dust- or sand-clouds are more easily seen with a polarizing filter in a radial direction. Sand-clouds can sometimes be very clearly observed on the beach, when a strong wind is blowing parallel to it and when one is looking into the distance along the beach. Even smoke can show polarization at 90° from the sun. Try having a look at it!

23. Polarization and negative polarization of clouds at twilight

After sunset clouds are no longer illuminated directly. Consequently, their colour changes from brilliant white to bluish grey, for the light is now coming from all sides of the blue polarized sky. The contrast of brightness between the sky and the clouds has diminished. The lower the clouds, the sooner this change occurs: high clouds are lit by the sun for a longer time (plate 18).

Differences in polarization between ice-clouds and water-clouds disappear at this time of day. The polarization of the blue light from the sky is taken over mainly by the clouds; so the polarization contrasts between the blue sky and the clouds (§ 16) almost disappear. In the area opposite to the sun, the clouds are negatively polarized, even more so than the negative polarization of the blue sky itself (§ 14). This is not surprising: under bright skies we see the light scattered from a long column of air, in which each particle may contribute to the negative polarization of the whole column. The size of the contribution depends on the distance of the particle from the observer. In the case of a cloud, the light is coming from a well-defined object at a given distance; it is the characteristic features of polarization of this 'single particle' that can be seen. Therefore, a cloud gives a better idea of negative polarization than can the blue sky itself (see §§ 14, 86 and fig. 20 on p. 32).

Try to find the neutral Arago point in these clouds (§ 14); it will usually be higher above the horizon than in the case of the blue sky. Such an observation will be most successful for instance with cumulus clouds, which sometimes cluster in beautiful banks.

24. Overcast sky and fog

Under a heavily overcast sky the illumination comes more or less from all sides and the polarization of the clouds largely disappears. Yet more light comes from the zenith, where the layer of clouds looks thinnest, than from the horizon; this is the counterpart of the distribution of light in an unclouded sky at twilight. Because of this preferential direction, the light of the clouds is *horizontally* polarized (§86). The degree of polarization is maximal just above the horizon but only amounts to 10–20%. This maximum degree of polarization is reached when the cloud cover is thick and the visibility is good. Higher in the sky, however, the degree of polarization decreases rapidly and higher than about 10° above the horizon it is negligible; in the zenith (the 'average source of light' at any solar elevation in this case) it must always be zero.

This polarization pattern can only be seen when the cloud layer is some kilometres thick, as during the passage of a weather front. Thinner clouds, like strato-cumulus or stratus do not show this horizontal polarization but are tangentially polarized with respect to the (invisible) sun (§21), as if their particles were still directly lit by the sun. Apparently, this light is still not sufficiently diffuse to bring about the above-mentioned pattern of polarization. The same situation occurs in fog, and the direction of the polarization enables us to determine the position of the invisible sun, as the Vikings did once before us. But when the clouds are very thick and the visibility is poor (for instance during drizzle, rain or snow), the illumination is extremely diffuse, so that it has hardly any preferential direction. Polarization is therefore absent in this case.

25. Noctilucent clouds and nacreous clouds

These two kinds of clouds will be discussed separately. They are only visible after sunset, and both are formed at extreme heights – the stratospheric nacreous clouds at about 25 km and the noctilucent clouds even up to 80 km. The latter occur in the mesosphere. All other clouds are tropos-

Plates 18–19 *Left:* after sunset the colour and the state of polarization of the clouds change (§23). *Right:* on a clear summer night the mesospheric noctilucent clouds can be frequently seen. Their light is polarized (§25).

pheric and therefore can never extend higher than 12 km above the Earth's surface.

The appearance of *noctilucent clouds* (plate 19) is often preceded by an exceptionally bright yellowish twilight in the direction of the already set sun when the sky is very clear. Then these undulating silky luminous clouds appear against the dark sky. Because of their great height they are lit by the sun for a long time after sunset. This finishes when the sun is more than 15° below the horizon. Their colour is bluish silver and they are usually low above the horizon in the same section of the sky as the set sun. They are typical summer clouds and are virtually only observable from May to August. Apparently, only in those months does the vertical structure of the upper atmosphere allow their formation.

They occur most frequently in northern latitudes; at latitude 52 °N they are visible on about 25% of the clear nights of July. The more beautiful complexes among them are of course rarer. Because of their great tenuity, these clouds are never visible at day.

The light of this form of clouds is tangentially polarized just like that of other clouds. Their polarization is, however, much stronger and may even reach a degree of polarization of as much as 96% at 90° from the sun. This is considerably stronger than that of blue sky. It is, nevertheless, not easy to see this polarization because of the very low light intensity of these mysterious clouds. Moreover, we hardly ever see them at angular distances greater than 50° from the sun (in which case the degree of polarization is still about 50%). The polarization of the light from the sky, however, hardly affects the situation because it is already so dark that the sky is emitting very little light. This kind of observation is best taken at as great an angular distance from the sun as possible; the nearer one looks at the sun, the lower the degree of polarization.

Unfortunately, the brilliant *nacreous clouds* do not occur in many countries. They appear on the lee of long mountain chains, when a strong wind is blowing from a direction perpendicular to the chain. In Norway and at the Antarctic these clouds are regularly seen; and in east Scotland they sometimes appear. They can be observed most beautifully when the sun has gone down but is still able to shine upon the stratosphere; then they appear at about 20° from the sun. Their riot of colour is characteristic of these clouds. They indeed show 'mother-of-pearl' shades. Their light is polarized, and according to Minnaert the colours change when the clouds are looked at through a polarizer and the filter is rotated. Such behaviour is to be expected when the cloud particles are about $0.8\,\mu$m in size. In this case, polarization of the scattered light is already considerable at a small angular distance from the sun and depends strongly on colour and angular distance (§86). Here also the dominant direction of polarization of the clouds is probably tangential with respect to the sun. On a journey to

Norway, one should not lose the opportunity to try to see this unique phenomenon. The also colourful *iridescent clouds*, which can be seen anywhere, are unpolarized (§43), and bear no relation to nacreous clouds.

Rainbows, haloes, glories and related phenomena

26. Introduction: optical phenomena in the sky

The phenomena that are dealt with in this chapter are very special, because of their great variety and wealth of colours. It is not surprising that the search for them is a special delight for every lover of Nature. Usually they appear in the sky as *spots*, *arcs* or *circles*; more often than not the last have the sun or its anti-solar point[1] as a centre. Many of these phenomena have colour shades like those of rainbows, and some have colours which are even more brilliant.

These so-called '*optical phenomena in the sky*' are caused by particles other than gas molecules in the air – water-drops and ice-crystals. They scatter the sunlight much more effectively than do the gas molecules, which are responsible for the blue light from the sky. The brightness of this light is so many times greater that only a small quantity of these particles in the atmosphere will suffice entirely to dominate the blue light from the sky. The aspect of the firmament quickly changes from blue to milky white when these particles are formed in the atmosphere; the observer experiences it as the appearance of a (thin) cloudiness.

It is no longer the scattering properties of the molecules, but the optical properties of water-drops or ice-crystals that determine the aspect of the sky. One can conceive that there are many ways along which a beam of light can travel in such a particle before leaving it. In some of them sifting of the colours of sunlight occurs, and this results in the appearance of *rainbows*, *haloes* and other colourful phenomena in the sky. In these cases sifted colours are apparently projected onto a particular section of the sky. The exact *nature* of the phenomenon which appears is of course determined by

[1] The anti-solar point is the point exactly opposite to the sun. Consequently, it is as far below the horizon as the sun is above the horizon, exactly on the place where one must imagine the eyes of one's shadow.

46

the kind of particles that are in the air; the optical properties of drops or ice-crystals are not identical. For instance, a rainbow will never appear in an ice-cloud but only in water-drops.

Polarization varies greatly between the individual phenomena. Some have strong polarization, usually tangentially directed with respect to the sun. This occurs, for instance when reflection near the Brewster angle (§74) plays a role in their formation. Other phenomena, however, display radial polarization or are barely polarized; the latter occurs for example when they are formed by diffraction or total reflection. But owing to their great variety, their occasionally surprising shapes, and their usually great richness of colours and of polarization, these optical phenomena are among the most delightful objects to be observed in Nature.

Fortunately, a number of these phenomena are much less exceptional than is generally supposed. This is, paradoxically enough, caused by their frequently high intensity of light, which often restrains us from looking at them with the naked eye. This holds particularly for those phenomena which are near the sun. It is usually much easier to make observations with sunglasses, which enable one to look directly. A few optical phenomena occur, however, so close to the source of light that one is forced to look almost straight into it. This is the case with the diffraction corona (§43), which can be observed more easily near the moon than near the sun.

27. Rainbows (plates 20–30)

These are the best-known optical phenomena in the sky. A rainbow is formed when the sun shines directly on drops of rain, and appears as a coloured band at about 138° from the sun, hence at 42° from the anti-solar point. At solar elevations higher than 42° the bow is entirely below the horizon and therefore invisible in the sky. The (primary) rainbow is red outside and blue inside; the innermost colours are usually somewhat paler than the red. On its inside, the colours sometimes recur once or more often: these are the so-called 'supernumerary bows'. Within the rainbow, the sky is plainly brighter than outside it. At about 8° outside the rainbow, a second bow can be seen to appear – the secondary rainbow. The latter is considerably fainter than the primary one. Its sequence of colours is reversed; outside this bow the sky is again somewhat brighter than within. Usually, only sections of these rainbows are seen, because the entire sky is not filled with drops or not all the drops are lit by the sun. On the other hand, the rainbow can sometimes be seen as a complete circle from a high tower!

Rainbows occur mostly in showery weather, hence under cumulonimbus clouds. But also after passage of a weather front they are sometimes seen appearing beautifully while the rain passes, at least if the front is

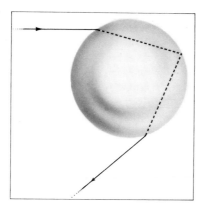

Fig. 24 The light path through a water drop which is responsible for the formation of the primary rainbow.

immediately followed by a sharp bright period. Moreover, they occur in waterfalls, geysers, sprinklers, fountains, flower-syringes, in sprays of sea waves, in bathroom showers, in drops of water in grass (the 'dew-bow', plate 22) or in dew-drops on cobwebs and in all other places where drops of water are directly lit up by the sun. There is of course no fundamental difference between all these bows: the colour sequence is always the same and they always appear at 42° from the anti-solar point; only in the case of drops of salt water does this distance prove to be about 1° less (plate 27). So, we must always take care to look in the right direction with the sun at our back. The colours of the rainbow are of course the same as those of which the source of light is composed; in the case of a red sunset, only the 'red rainbow' can be seen (plate 23).

We can also see rainbows generated by artificial light or by moonlight; certainly in the latter case the bow seems white as a result of the poor colour sensitivity of our eyes when the light intensity is low. Once one was even seen during a total eclipse, in which the solar corona functioned as source of light! At the time the colours were pink and green. These are the colours of the brightest emission-lines of the solar corona.

However the rainbow may come about, its light always has a very high tangential polarization. The degree of polarization of both the primary and the secondary rainbows is so great that they can be completely 'rotated away' with a polarizing filter (plates 24–26). But with the filter in the tangential position, the bows appear very clearly. Indeed we can then often see the secondary rainbow, when it is not yet visible to the naked eye. Also it is much easier to find faint rainbows, like the red one or the dew-bow, with a polarizing filter.

The strong polarization is the result of the path that the beams of light generating the rainbow must follow through the drops. In the case of the primary rainbow, the beams are always subjected to one internal reflection against the back of the drop (fig. 24); this reflection occurs twice in the case of the secondary rainbow. The reflections happen to occur very close to the Brewster angle (§ 74) so that the light is very strongly polarized. In contrast to what happens in a single external reflection, this light suffers a change in its direction much greater than 90° because of additional refractions (fig. 24). As a result, the strongly polarized phenomenon appears much further than 90° from the sun (see also § 75). The degree of polarization of the secondary rainbow turns out to be slightly less strong than that of the primary rainbow (90 % as against 96 %). In practice, however, it only means that both rainbows disappear entirely when a radially directed polarizing filter is used.

So, the very high polarization of the light of the rainbow leads to a striking contrast: with a tangentially directed filter both rainbows are brilliantly clear, but with a radially directed filter they disappear completely.

28. The sky near rainbows

Inside the primary rainbow and outside the secondary one the sky gives distinctly more light than in between them. Most light comes from the section within the primary rainbow. During rain, the area between the rainbows is by far the darkest section of the sky (plate 20). This dark band is called the *Alexander's band*, named after its discoverer, Alexander of Aphrodisius (A.D. 200). The extra brightness of the sky inside the rainbow (or outside the secondary rainbow) is caused by a path of light through the drop, which is similar to the one which gives rise to the adjacent rainbow. In fact the rainbow is only the *boundary-line* of a bright area in the sky, lit up in rain as a result of a specific combination of refractions and reflections, and is therefore not a solitary phenomenon (see plate 21). This boundary-line varies from colour to colour and red is its outermost limit. The brightness in such an illuminated area is greater the closer one looks in the direction of the boundary-line, so that the latter appears as a bright vividly coloured band known as the rainbow.

Since the light within the primary rainbow (or outside the secondary one) comes about in the same way as that of the rainbow itself, its polarization is the same also. So, with a radially directed filter, not only does the rainbow disappear but so does the extra brightness at its inner side (plates 24–26). The Alexander's band can no longer be seen, because the whole area where the rainbows have been extinguished now shows the same dark, uniformly grey light. In fact, we now see what the distribution of light in the firmament would be like, if the drops of water were black and not transparent. Without a filter the same holds for the distribution of intensity in the dark Alexander's band, because the small amount of light coming

Plate 20 A double rainbow. The bows have an opposite sequence of colours. Between the bows the sky is very dark: the Alexander's band (§28). At the inner side of the primary rainbow (the brightest of the two bows), a supernumerary rainbow is visible. (Photograph G. Doeksen.)

from there arises only from *external* reflections upon the drops. On closer inspection, this light also turns out to be tangentially polarized. The effect, however, pales entirely into insignificance besides that of the bright adjacent rainbows.

During a shower of rain, the remaining part of the sky also has a mainly tangential polarization, with its highest degree at about 80° from the sun, where the external reflection on the drops occurs at the Brewster angle. However, this polarization, the degree of which is comparable to that of the clouds, is not as visually striking as that of the rainbows. Closer than 80° to the sun the drops light up chiefly as a consequence of refractions within themselves. Accordingly, the sky is much brighter here than in the area opposite to the sun. There is almost no polarization: as far as it does occur,

Plate 21 A complete rainbow, a wide-angle view. Inside the bow the sky is brighter than outside (§§27 and 28).

it is very weak and radially directed, as is to be expected from refracted light.

29. Reflection-bows and abnormal rainbows

Rainbows are sometimes abnormally shaped. The reflection of a rainbow (a *reflected bow*) may be seen in a smooth water surface, and the reflection of the sun is able to generate a rainbow in the sky. The latter is called the *reflection-bow* or the reflected-light rainbow (plate 28).

When a rainbow is reflected, the polarization of its light hardly changes: the plane of polarization is not reflected in the case of such a grazing reflection (§87). Therefore, contrary to the situation with the non-reflected bow, the direction of polarization of the reflected one no longer follows the curvature of the bow. Hence tangential polarization is now absent. But the rainbow and its reflection can be extinguished almost simultaneously

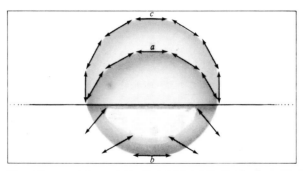

Fig. 25 $a=$ rainbow; $b=$ reflected rainbow; $c=$ reflection-bow. Arrows indicate the direction of polarization.

with a polarizing filter in the appropriate orientation (fig. 25). Of course, a reflection of the rainbow can also be observed by putting a polished metal plate or an ordinary mirror on the ground. Then, the direction of polarization does follow more or less the curvature of the bow, because these materials reflect not only the bow but also mirror the plane of polarization (plate 52 on p. 83 and §91). In these artificial circumstances it is the metal plate in particular that additionally converts a part of the linear rainbow-light into circularly polarized light: the right part of the reflected rainbow is then left-handed circularly polarized and vice versa (§91). When a horizontal glass plate or a pane is used as a reflecting surface, the polarization is analogous to the one seen in reflection on water.

Reflection-bows are very rare and usually only sections of them are seen. The polarization of this type of bow is tangentially directed with respect to the *reflection* of the sun. Hence the polarization follows the curvature of the bow (see fig. 25). Some deviations may occur, because the light of the reflected sun is itself horizontally polarized, but generally they will be hardly perceptible. The *reflected dew-bow* and the *reflection dew-bow*, which can be seen in ditches, behave in the same way as the reflected rainbow and the reflection-bow so that by finding out whether the polarization follows the curvature of the bow, one can decide which of the two types of bow is seen here.

Under extremely rare circumstances extra bows sometimes appear in the sky in different places. It usually happens near water surfaces. Some of them are reflection-bows but come from reflection on rough water; other ones, however, can not yet be explained satisfactorily. If you should be lucky enough to see such a reflection-bow or another abnormal rainbow, take a picture and examine its polarization. Does the polarization direction follow the shape of the bow or not? Is the degree of polarization higher or lower than that of the rainbow itself? Observation of the polarization of this kind of phenomenon may hold the key to the explanation of such remarkable and rare bows.

30. Fog-bows and cloud-bows

As no two trees are alike, so one rainbow differs from another. The greatest contrast exists between a magnificently coloured rainbow, as seen in a passing shower, and the dull white fog-bow, which is formed in minute drops of water.

The outward appearance of the rainbow changes as the drops of water become smaller and this can easily be established with a flower-syringe (plate 29). The colours begin to overlap each other and the bow turns whitish with only a faint colouration along the edge. Its radius somewhat decreases and the supernumerary bows are farther away from the bow. It is usually more difficult to see secondary rainbows. Very small drops, as occur for instance in clouds, show the rainbow only as a slightly intensified brightness of the light at about 145° from the sun; the colours have completely disappeared. Only an experienced observer is able to establish the presence of the bow in clouds (plate 30).

The tangential polarization of this *cloud-bow* is, however, still present, although it is somewhat weaker than in the case of big drops. The cloud-bow betrays its presence by this polarization: with a polarizing filter we can see that the·light of sunlit water-clouds at about 145° from the sun is polarized to an abnormally high degree, certainly in comparison with ice-clouds! This has already been mentioned in §21. So, with the aid of a polarizing filter, everybody can easily observe the cloud-bow. Strange to relate, rainbows can be seen almost every day, though merely in the form of the inconspicuous cloud-bow.

In *fog*, the shape of the bow usually remains more easily recognizable, even frequently showing some red colourization at its outer edge so that without the aid of a polarizing filter the fog-bow can easily be found. It can also be observed in the steam of a bathroom shower when a thin ray of sunlight enters the room. Then, the bow is seen as a 'thickening' of this ray of light, which means a greater local brightness.

If the tangentially polarized fog-bow appears, the radially polarized rings of the glory are usually visible near the observer's shadow. In Nature or in the bathroom, the simultaneous appearance of these oppositely polarized phenomena yields interesting observations (see §42).

31. Supernumerary fog-bows

The polarization of fog-bows is less than that of the rainbow so this bow cannot be totally extinguished with a polarizing filter. Then something odd appears to happen to the fog-bow if observed with a polarizing filter: not only do the supernumerary fog-bows become partly extinguished when the filter is in a radial direction but they also appear to be *shifted* from the

Plates 22–23 *Left:* dew-bow on grass. *Right:* a red rainbow at sunset (§27).

Plates 24–26 Polarization of a rainbow. *Below:* without polarizing filter. *Top:* with a polarizing filter in front of the camera in two positions (§27).

Plates 27–28 *Left:* above the horizon, the rainbow is visible in freshwater drops, below the horizon in drops of salt seawater spray. The latter rainbow is one degree closer to the antisolar point. (Photograph J. Dijkema.) *Right:* part of the reflection-bow appears between the bows (§29). (Photograph K. Lenggenhager.)

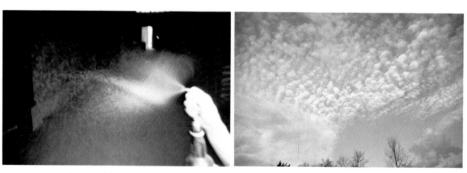

Plates 29–30 *Left:* rainbow in flower-syringe. The smallest drops (close to the syringe) generate a paler bow. *Right:* the extra brightness of the clouds in the centre of the picture is the result of rainbow-scattering (§§21 and 30).

positions they hold when observed with the filter in a tangential direction. They are now exactly between the supernumerary bows in the bright (tangential) direction. This strange effect is caused by the extra phase-shift of 180° in vertically polarized light, which occurs at reflection when the angle of incidence passes the Brewster angle (see §87) and is also responsible for the mirroring of the polarization plane by reflection of light which has a less oblique incidence than the Brewster angle. Supernumerary bows are the result of interference between two rays of light that have followed different paths through the drop but which have both been reflected once against the back of the drop (fig. 26). When the one ray is reflected with an angle of incidence smaller and the other with an angle of incidence larger than the Brewster angle, there is an extra phase difference of 180° for vertically polarized light. Therefore, interferences that would otherwise be constructive become destructive (§87). This causes a shift of the supernumerary bows arising from vertically polarized light. In the case of a rainbow, the internal reflection occurs so close to the Brewster angle that the shift still happens when the diameter of the drop is as much as

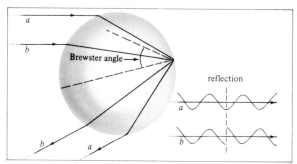

Fig. 26 Formation of supernumerary bows by interference of rays of light (*a*) and (*b*). For vertically polarized light, (*b*) is subjected to an extra phase-shift, which does not occur in the cases of (*a*) and horizontal light. This causes an altered position of the supernumerary bow in radially polarized light.

about 1 mm. But it is only when the drops are much smaller and as a result the degree of polarization has greatly diminished, that we can observe this bow with a radially directed filter.

The observation is rather difficult because supernumerary fog-bows are fairly rarely seen by Earth-bound observers. This means the challenge is all the greater to observe the phenomenon in the open country.

32. A thirteenth rainbow?

Rays of light that leave a drop of rain after three, four or more internal reflections may equally well generate a rainbow. These '*higher-order rainbows*' appear mostly in a different section of the sky from where the primary or the secondary rainbows are seen. Apart from one curious exception, mentioned in §42, they are never observed outside the laboratory. The reasons for this are that their light intensity has been too much reduced by all these reflections and that they are completely dominated by light that otherwise lights up the sky via the drop. It is, however, easy to show their existence by studying a glass ball filled with water and lit by the sun: at certain angles we can see these rainbows as colourful points appearing at the rim of the ball. Then it is easy to establish that the light of these rainbows is also tangentially polarized. In Nature, however, only the primary and secondary rainbows are observable, being rather bright and appearing in a dark section of the sky.

Nonetheless, it has been suggested that the *thirteenth rainbow* (thirteen reflections against the wall of a drop!) should also be visible under favourable natural circumstances with the aid of a polarizing filter. This bow should appear at about 80° from the sun. During rain the sky is rather dark and quite strongly tangentially polarized in this area. Theoretically, the polarization of this background light may be considerably stronger than that of the thirteenth rainbow, which amounts to about 75% and is also tangentially polarized. With a radially directed polarizing filter, the background light would, therefore, be more readily extinguished than that of the rainbow and theoretically speaking the latter should be apparent. However, it does not seem very likely that such an observation could be concluded successfully in natural circumstances so that this bow can really only be studied under laboratory conditions. Of course, in rainy weather one can still try to perceive this dim rainbow with a polarizing filter, however improbable may be the outcome.

33. Haloes: their shape and appearance (plates 31–40)

Rainbows and glories (§42) are optical phenomena which can only appear if drops of water are present in the atmosphere (or on the ground). All these phenomena become impossible when the drops freeze, forming snow- or ice-crystals. But transparent ice-crystals will also refract and reflect sunlight, as a result of which various circles, spots or arcs appear in the sky, and a number of these sometimes display beautiful rainbow-like colours. Such circles usually have the sun (or moon) as a central point, as the rainbow always has the anti-solar point as a central point. Optical phenomena caused by refractions in ice-crystals usually appear on the same side of the sky as the sun. Collectively, they are called *haloes*. They are not at all scarce: a halo can be observed on an average of two out of three days somewhere in the Netherlands. But in contrast to the more exceptional rainbow relatively few people have seen them. The rainbow, however, appears in the darker section of the sky opposite to the sun, whereas most haloes are visible at a relatively short distance from the sun. For this reason, their intensity of light is so high that it is often difficult to look into them with the naked eye. But with sunglasses these beautiful phenomena can be easily observed, frequently when there are cirrus-like clouds near the sun.

There exists a great variety of types of halo, because ice-crystals can refract or reflect sunlight in many different ways, but the most important and frequently occurring haloes can be divided roughly into three groups:

1 Light-intensive coloured phenomena at about 22° from the sun (or from the moon). The form can be a circle with the sun at its centre (the *22° halo*), but it can also be very bright spots on either side of the sun (the *parhelia*) or curved arcs of various shapes above or below it. Red is nearest the sun and usually very conspicuous. Orange and yellow are usually still distinctly visible, but the other colours are paler. This category of halo occurs most frequently.
2 Fainter but more colourful phenomena of the same character and with the same sequence of colours at about 46° from the sun. Their abundance of colours may exceed that of rainbows.
3 Uncoloured (white) spots and bands in several places in the sky. Here haloes belonging to this group will be called *reflection haloes*.

By moonlight nearly all haloes look colourless, because under those conditions our eyes are almost insensitive to colours. It is only the bright parhelion (at night, called *paraselene* or *mock-moon*, plate 34) that still appears to display colours. Figure 27 is a sketch of the most important halo phenomena. The haloes are certainly worthy of observation, not only for their beautiful colours and variety but also because they may alternately appear or disappear very quickly. Some sub-horizon haloes have also been included in fig. 27; generally, they are only visible from aeroplanes or from high mountains. The most frequent varieties are the *22° halo* (*c*), the *parhelia* (*mock-suns*) at 22° on either side of the sun (*a*), the *upper and lower tangent arc* to the 22° halo (*d* and *e*), the *circumzenithal arc* at 46° above the sun (*b*) and the *sun pillar* (*i*). The *46° halo* (*f*), the *parhelic circle* (*g*) and the *paranthelia* (*h*) are much more uncommon.

Special attention should be paid to the lower tangent arc to the 22° halo (fig. 27, *e*), because at solar elevations between 10° and 13° it suddenly transforms its shape into a very remarkable loop. The intersection of the loop is white and it is always exactly in the position where the sub-sun (§39)

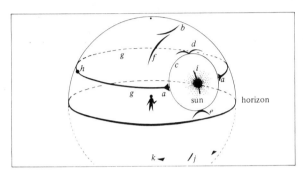

Fig. 27 Sketch of the most important haloes: parhelion (*a*), circumzenithal arc (*b*), 22° halo (*c*), upper and lower tangent arc (*d* and *e*), part of the 46° halo (*f*), parhelic circle (*g*), paranthelion (*h*) and the sun pillar (*i*). The subsun (*j*) and subparhelion (*k*) are haloes below the horizon. Haloes (*g*), (*h*), (*i*) and (*j*) are uncoloured.

Plates 31–32 *Left:* the 22° halo. *Right:* tangent arcs and parhelia at the same distance from the sun are formed, if ice-crystals have a preferential orientation and no longer whirl randomly through the air (§34). Wide-angle pictures.

Plates 33–34 *Left:* parhelion (mock-sun) at 22° from the sun. *Right:* paraselene (mock-moon), the same phenomenon at 22° from the moon. The bright point near the mock-moon is the planet Jupiter.

Plate 35 Under the parhelion a sub-parhelion is formed (§39). It is always below the horizon, often visible from aeroplanes.

should appear. At these solar elevations the arc changes its outward appearance so quickly that one runs the risk of being taken by surprise; it transforms its shape from minute to minute. Being, however, completely below the horizon, the arc cannot be seen from the ground, but only from an aeroplane or from a high mountain. Also, the very rare lower tangent arc to the 46° halo behaves in exactly the same way, at solar elevations between 22° and 28°, by turning into a loop with an intersection in the position of the sub-sun. I do not know if it has ever been observed.

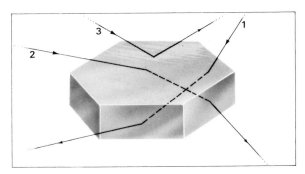

Fig. 28 Light can follow many paths through an ice-crystal; each path causes the formation of another halo.

34. Haloes: how they come about

There are so many kinds of haloes that it is useful to examine how they come about, before talking about their polarization effects.

All haloes are formed by refractions or reflections on ice-crystals floating in the atmosphere. These crystals are generally hexagons with flattened ends. As in a glass prism or a polished diamond, in such a crystal many combinations of refractions and reflections are possible and can result in a great number of paths of light through it. Some of these paths have been sketched in fig. 28. Which halo is formed in the sky depends upon the path that the light has followed through the crystal. In addition, the orientation of the crystal in the atmosphere is also important: different preferential orientations of the crystals result in the generation of different haloes within a given halo group. When the orientation is random, and the crystals thus whirl like the leaves of a tree through the atmosphere, round coloured circles appear round the sun; when all the crystals have taken up the same orientation, the more vividly coloured spots or arcs are formed instead at about the same distance from the sun as were the circles (plates 31–32). Light-path 1 in fig. 28 is responsible for the formation of all halo phenomena in the first group, i.e. at 22° from the sun. So which member of this group will appear depends only on the orientation of the crystals. The same path of light is depicted in fig. 29; the emergent light is twice refracted during this path by faces forming an angle of 60° with each other. Haloes formed in this manner occur most frequently.

When light is refracted by crystal faces forming an angle of 90° with each other, as in the case of light path 2 in fig. 28, members of the rarer group of haloes at 46° from the sun are formed. Reflections against the faces of ice-crystals cause the haloes of the third group (fig. 30). These reflections can occur externally or internally, and in the latter case can also be total; then the intensity of reflected light is many times greater than in a non-total reflection.

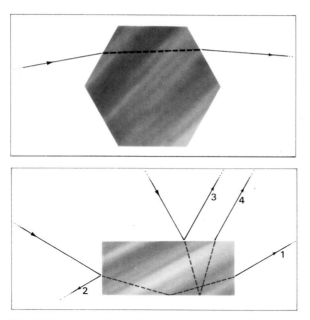

Fig. 29 Passage of rays through ice-crystals, which is responsible for the formation of haloes at 22° from the sun (ray 1 in fig. 28).

Fig. 30 Internal and external reflections also may cause halo formation. Reflection 1 is total; 2, 3 and 4 are not.

The colours displayed by a large number of haloes are the result of the sifting of the colours of sunlight. Exactly as in the case of a rainbow, such a colourful phenomenon is strictly speaking nothing but the definite boundary of a particular area in the sky which is lit up by a specific combination of refractions and reflections in ice-crystals. Here too the outer boundary limit of this area is formed by red light. As this limit in practically all coloured haloes is on the inner side of the halo (the part closest to the sun), the sky inside the halo is yet darker than that outside.

35. Polarization effects of haloes: a general view

Research into these effects is an almost untrodden area. Yet these phenomena can evidently show beautiful and unexpected polarization effects: the polarization of halo light can be as varied as the types of the haloes themselves. Since, however, there is no essential difference in the light paths which lead to formation of any member of a certain group of haloes, no distinct difference in polarization exists among such haloes. That is why the 22° halo, the parhelion or the tangent arc to the 22° halo show basically the same polarization. Between haloes of different groups (e.g. at 22° and 46° from the sun) these differences do exist, because they are formed by completely different light-paths through the ice-crystal. If we want to understand from where the polarization of a certain halo comes, we must know precisely which refractions and reflections gave rise to it.

Haloes in which only refractions, not reflections, are playing a part, are *radially* polarized. Thus this polarization is *stronger* when the refraction is stronger, i.e. according to the distance that the phenomenon is situated from the sun. So, the haloes at 22° from the sun are less polarized than members of the 46° halo group. In fact, only the polarization of the latter group is strong enough to be observed. Forms of haloes in which reflections play a part, can be more strongly polarized; the direction is then *tangential* with respect to the sun. This strong polarization only occurs if the reflection is not total or if the quantity of light which we receive via total reflection is small compared to the non-totally reflected part. However, in this case the polarization of these haloes is not always very conspicuous, because its direction is the same as that of the polarized blue sky or the clouds against which they are outlined, and the maximum polarization usually occurs at about 90° from the sun. Hence, through a filter there is little change of contrast between the halo and its background: both are weakened by about the same amount.

Furthermore, it is important that ice-crystals are doubly refractive (birefringent). This leads to curious, but little known, effects of polarization for some haloes like those at 22° and 46° from the sun. Ice-crystals are uniaxial, positive, hexagonal crystals, the optical axis being perpendicular to the ends (see fig. 31 on p. 64). A beam of light, which travels in a direction perpendicular to this axis is maximally subject to this double refraction; among the haloes, this particularly applies to the above group at 22° from the sun.

From the foregoing it may appear that the number of polarization effects of haloes is large and specifically dependent on the generating light-path through the ice-crystal. How some (rare) forms of halo come about is still unknown. Observation of their polarization may give some clues as to their origins, because this additional information may exclude some proposed mechanisms. A study of haloes with *circularly* polarizing filters, however, will hardly bear fruit because there is very unlikely to be a perceptible circularly polarized component in the light in the particular halo under scrutiny.

Finally, the following should be noted. Contrary to the situation with the circles, the direction of polarization of haloes like the tangent arcs, may not be exactly radial everywhere because of the preferential orientation of the generating ice-crystals. Parhelia and similar phenomena can also show a slightly altered direction of the plane of polarization. However, differences of this kind, which are the result of the orientation of the crystals, are generally so small, that it usually suffices to refer only to the direction of polarization of the circles (i.e. radial or tangential).

Plates 36–37 *Left:* parhelion, generated with a quartz crystal. The lower part of the photo has been covered by a polarizing filter; here the right section of the parhelion has been extinguished. Rotating the filter a quarter of a turn will cause this right section to reappear and the left section to disappear. This also occurs with natural parhelia (*right*) but here the polarized components are seven times nearer to each other and overlap to a greater extent (§36).

36. Polarization of parhelia, circles and arcs at 22° from the sun

The group of haloes at 22° from the sun (the 22° halo, parhelia, tangent arcs (plates 31–34 and fig. 27c, a, d and e)) are all formed by a weak refraction and, because of this, hardly any polarization may be expected. It can be calculated that the degree of polarization must be a mere 4%. This is too weak to be observed. When we study these phenomena through a polarizing filter, however, a very remarkable effect is evident: the haloes appear about 0.11° *closer* to the sun for radial vibrations than for tangential vibrations! This holds for all members of the 22° halo group, but is most striking for bright *parhelia* (plate 37), because generally they have a clearly defined red edge on the side towards the sun. So, a parhelion cannot be extinguished with a polarizing filter as for example can rainbows, but the whole spectacle *shifts* its position when we rotate the filter before the naked eye. This shifting of 0.11° does not seem very much, but it is clearly perceptible: it is about a quarter of the apparent diameter of the moon and the cirrus clouds usually act as a good reference point from which the shift can be observed.

This curious phenomenon is caused by the doubly-refracting properties of the ice-crystals in which haloes are formed (§83). On its way through the crystal, unpolarized sunlight is split into two rays which are perpendicularly polarized with respect to each other. Each of them generates its own halo (fig. 31). These polarized haloes have shifted only 0.11° with respect to each other, hence mutually overlapping to a great extent (the distance between the red edge and the yellow part of a 22° halo is about 0.5°), but with a polarizing filter these haloes can be seen individually. The

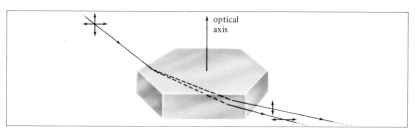

Fig. 31 As a result of double refraction, unpolarized light, passing through an ice-crystal, is split into two polarized rays, each of which generates its own halo.

shifting, which can be observed with such a filter, is the same for all colours. The most weakly refracted ray vibrates perpendicularly to the optical axis of the crystal so that the 'radial' halo is closest to the sun. As for the parhelion, the polarization of the component nearest to the sun is actually *horizontal* instead of radial, because here all the ice-crystals are oriented in such a way that their optical axis is vertical to the horizon. So, the outer red edge of the parhelion is 100% horizontally polarized; elsewhere horizontally polarized light is mixed with vertically polarized light of a somewhat different colour. This mixture can be neutralized with a filter, by means of which the colours of the parhelion become somewhat more clearly defined. In fig. 32 the polarization of the parhelion has been shown schematically.

The shift of parhelia, as depicted above, can be easily imitated in a graphic manner in the living-room, with a quartz crystal (plate 36). By causing sunlight to enter the crystal, a 'quartz-parhelion' can be created on a screen. As quartz has a double-refraction about seven times stronger than ice, the 'parhelia' for the two directions of polarization are shifted to a greater degree than in Nature: the red of the vertically polarized component coincides more or less with the blue of the horizontal component. If a polarizing filter is put in the beam of light, one of the components can be extinguished. It is noticeable then, that some colours of this parhelion become much deeper because they are no longer mixed. With the other polarized component, rotation of the filter now gives a shift of about 0.7°, thus admirably demonstrating this remarkable effect (see also §51).

At a high solar elevation (above 45°), the natural parhelion as a whole may show some polarization, because then the refractions are stronger. You should try to watch it. The direction of this polarization can deviate a little from horizontal, because of internal total reflections of the light in the crystals. Unfortunately, parhelia are rather rare at such high solar elevations and the polarization is not nearly so spectacular as the above-mentioned shift; especially as at best the degree of polarization does not exceed 20%.

It has been suggested that some parhelia, particularly at high solar

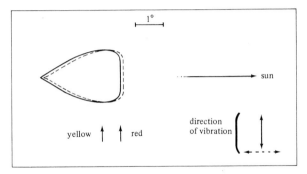

Fig. 32 The parhelion for two directions of polarization. Without a polarizing filter we see them overlap each other; with a polarizing filter we can see either of them individually.

elevations, are not generated by direct sunlight but by light of the sub-sun. Thus they may be termed the *sub-parhelia* of the *sub-sun* (§39), and would be expected to show a strong horizontal polarization similar to that of the sub-sun. It is doubtful, however, whether parhelia are often formed in such a way. Observations with a polarizing filter may prove their existence.

37. Haloes at 46° from the sun

This group includes the 46° halo (fig. 27f) and the circumzenithal arc (fig. 27b), which occur vertically above the sun. These haloes are less bright than the ones at 22° from the sun, but they surpass them in wealth of colours. Without doubt, the finest is the frequently very beautiful *circumzenithal arc* (plate 38), which often appears together with the parhelia when the solar elevation is not too high ($\pm$ 15–25°) (see fig. 33). These 46° haloes are formed by stronger refraction than the 22° ones, and their radial polarization is clearly visible. Their degree of polarization is 15–20%. Since the blue sky and the clouds at 46° from the sun are also quite polarized, but

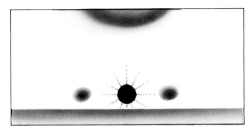

Fig. 33 Typical halo-sky at low solar elevation. When a bright parhelion appears to the left or the right, it is very likely that high above the sun the circumzenithal arc will also be visible.

in a reversed direction, we see a very distinct increase of contrast between halo and its background by watching it with a polarizing filter held in a radial direction; in the other direction the halo is considerably extinguished with respect to the background.

In addition, the haloes at 46° also show double-refraction effects, similar to the ones described in the preceding section. Hence, they also shift their position when seen through a polarizing filter which is then rotated. However, I know from experience that the effect is usually not as easily observable as in the case of parhelia. The reasons are that in general the light intensity of the 46° halo group is not very high and that the polarization of the halo as a whole and of its background may obscure it. However, in exceptionally bright haloes – some circumzenithal arcs for example – the effect is very clearly defined! The shift is about 50% larger than that of the parhelion but, as the colour-band is now three times broader, the result is relatively slightly less conspicuous than for the parhelion. It is interesting that now the *tangential* halo is nearest to the sun, hence exactly the reverse of the situation with the 22° halo group. So, with a tangentially directed filter we can see two things simultaneously: the halo is distinctly weakened and it is nearest to the sun.

38. Circles and tangent arcs at unusual distances from the sun

These are rare halo phenomena appearing at distances other than 22° or 46° from the sun. They are formed by refractions through ice-crystals of differing shapes. The polarization of these phenomena will generally be low, because their radius is also usually relatively small. We may, however, expect double-refraction effects to occur in some of these haloes. When you catch sight of such a halo-form, just estimate its radius and try measuring the shift seen through a polarizing filter perhaps with the aid of a pair of binoculars held behind it. In which direction has the red inner side been polarized? This last piece of information, together with an estimation of the radius, is often sufficient for interpretation of the phenomenon; indeed it is better than merely measuring the radius of the halo as accurately as possible.

39. Reflection haloes

The silvery-white *sub-sun* (fig. 27*j* and plate 40) can frequently be seen from aeroplanes and even from spacecraft (plate 8 on p. 21); it is probably the most frequent halo-form seen from these elevated positions. It often appears as an oblong spot, situated exactly below the sun in the place where one would expect the reflection of the sun on the Earth's surface. It is indeed a reflection of the sun; the reflecting face is, however, not a water surface but the horizontal faces of many oriented ice-crystals, and yields light via the paths shown in fig. 30. The phenomenon is always below the horizon and often outlined against the clouds under the aeroplane. The sub-sun usually shows a strong, horizontal polarization which depends on the solar

Plates 38–39 *Left:* circumzenithal arc at 46° above the sun, often appearing together with parhelia and showing remarkable polarization effects (§37). (Photograph K. P. Bijleveld.) *Right:* Parhelic circle with paranthelion; they belong to the reflection haloes (§39). (Photograph P. P. Hattinga Verschure.)

Plate 40 Sub-sun, a strongly polarized reflection halo; often observable from aeroplanes (§39). (Photograph G. J. Heinen.)

elevation. It may also happen, however, that so much totally reflected light contributes to its formation that there is hardly any polarization. In this case the sub-sun is chiefly formed by light-path 1 in fig. 30. Sometimes a coloured *22° halo* appears around the sub-sun; it must show the same horizontal polarization. To the left and to the right of the sub-sun coloured parhelia are often seen: the *sub-parhelia* (fig. 27k and plate 35). They appear, therefore, exactly under the ordinary parhelia. If they are parhelia of the sub-sun or a reflection of the usual parhelia on the horizontal faces of other ice-crystals, their polarization must be horizontal and as strong as the light of the sub-sun. They may, however, also (and probably more often) be formed by total reflections within the generating ice-crystal; in which case they are unpolarized and may, indeed, be brighter than the sub-sun itself. In all cases, they must also show the double-refraction shift of 0.11° similar to that of the ordinary parhelia (§36). Sometimes the milky white *parhelic circle* and the *paranthelia* (fig. 27g, h and plate 39) appear above the horizon at the same elevation as the sun. The parhelic circle has the zenith as a central point and passes through the sun; the paranthelia usually appear as thickenings in this ring at a distance of 120° in azimuth from the sun. Both phenomena show a tangential polarization but usually to a lesser degree than that of the clouds against which they are outlined. A higher degree of polarization would be expected from generation via light-path 2 in fig. 30 (external reflections against the vertical faces of oriented ice-crystals), but evidently total reflection often plays a greater part. Is that always the case? Not according to some explanations, which suggest that the polarization at 90° from the sun could be very high. The polarization is in any case not spectacular, the more so because its direction is equal to that of the blue sky and the clouds; so through a polarizing filter we see hardly any changes of contrast. The white *sun pillar* (fig. 27i) is formed by reflections against rotating or vibrating ice-crystals. Yet its light is only slightly polarized, except perhaps at greater distances from the sun. Here, the polarization must be horizontal, just as in the case of the sub-sun. When the sun is very low, the pillar appears most often as a long vertical band passing transversely through the sun. When the sun is red, the pillar naturally shows the same colour.

40. Haloes on blurred panes

Minnaert says that when very cold panes are blurred, for example by breathing against them, ice-crystals will be formed upon them which may result in the formation of haloes. These can be seen around a source of light observed through the pane. We do not see a diffraction corona (§43) as usually seen in blurred panes, but a halo with a radius of about 8°. This halo seems double and it has been suggested that it could be caused by double-refraction. In that case these components would have to be polarized in the same way as are parhelia. It is, therefore, possible to decide with a polarizing filter if this explanation is indeed the right one, although it seems unlikely.

41. Artificial haloes

By crystallizing a saturated salt-solution (e.g. by cooling or by adding alcohol) salt-crystals that show haloes are formed in the solution, usually with a radius of about 10°. To observe them, one pours the solution into a flat bottle, keeps it close to the eyes, and looks at the sun or another source of light. This experiment is very successful with alum. These haloes are not polarized, but an artificial halo made in this way in sodium nitrate, will be very strongly radially polarized. The reason is that sodium nitrate is doubly-refractive, and this double-refraction is 175 times stronger than that of ice. The split of light in polarized beams is, therefore, so large that radial and tangential haloes are completely separated and are, consequently, at a great distance from each other. In a solution, the tangential haloes even completely disappear, since the refractive index for this polarization direction is almost equal to that of the solution which contains the crystals.

42. The glory (plate 41)

This is perhaps the most remarkable phenomenon in atmospheric optics. It consists of a coloured aureole that is visible around the shadow of the observer's head. As he has never seen the aureole appear around other people's heads, the unwitting observer may therefore suppose that the aureole is an indication of his being superior to other people! Unfortunately, the fact that others may just as easily observe such an aureole around their own heads, more or less obviates this explanation.

The glory only appears when the shadow falls upon small water drops, such as those in clouds or fog. So, the glory is not a halo. One must be above the fog or clouds in order to see it, which is a rather exceptional situation. Therefore, observations from times past are rather rare. Nowadays the opportunity to see a complete glory is greater, because many people fly by plane above the clouds. However, parts of the glory can also be observed in terrestrial circumstances if one only knows where to look and can achieve the proper situation.

The size of the glory (some degrees in diameter) depends only on the size of the drops in which it is formed and not upon the distance of the observer from the cloud. The smaller the drops, the larger the glory. On the other hand, the size of one's shadow does, of course, depend on the distance from the cloud: so when one recedes from the cloud, the shadow grows smaller but the glory keeps the same dimension. The centre of the glory is exactly where the shadow of one's eyes might be and thus precisely opposite to the sun. In former days pilots used to avail themselves of this property for navigation: instead of 'shooting' the sun, they sometimes 'shot a glory'. Its depth below the horizon corresponds exactly to the desired solar elevation. When flying with the sun at the back, this kind of measurement is more convenient. When one walks through the aeroplane to its front part, it becomes evident how exactly the glory is standing opposite to the sun: the glory is moving with you along the shadow of the aeroplane. Of course everyone on board sees his own glory.

On closer inspection the glory appears like this: a bright white area directly around the shadow, with a red edge, surrounded by some coloured rings with blue inside and red outside. The innermost red edge is brightest. Although becoming progressively fainter, the colours may repeat themselves up to five times. Nearer the Earth's surface the glory can be observed

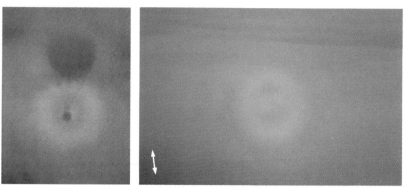

Plates 41–42 *Left:* glory around the shadow of the basket of a balloon. (Photograph A. G. F. Kip.) *Right:* glory, seen from an aeroplane on remote clouds with a polarizing filter. (Photograph A. B. Fraser.)

upon low fogs, most successfully when the background is dark (e.g. over an asphalt road). Parts of the glory are also visible in the vapour above a cup of tea, in the 'cloud' breathed out on a cold day, or in the vapour of a bathroom shower. In the last case it is best observed when a thin beam of sunlight is shining into the bathroom. The phenomenon can often be visible from mountains or high towers rising above a layer of fog. The glory is usually seen together with a fog-bow (§30).

The polarization of the glory is very striking. The coloured rings are *radially* polarized, hence contrary to the polarization of the fog-bow. This contrast can be clearly observed in the steam of a bathroom shower. However, close to its centre, in the white area, the glory is *tangentially* polarized; the central point itself has no polarization. This curious contrast is best seen when one can observe a bright glory in its entirety. With a vertically directed polarizing filter the upper and lower sides of the coloured rings become brighter and they look thicker. On the other hand, in the white area close to the centre, again on the upper and lower sides,

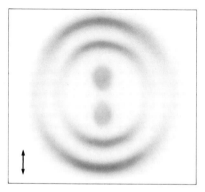

Fig. 34 The glory seen through a filter transmitting vertically polarized light. The innermost structures are dark and bluish.

dark bluish wedge-like triangular spots appear which point to the centre of the glory (fig. 34 and plate 42). Obviously, when the filter is rotated, this structure co-rotates. The triangular spots resulting from the tangential polarization are often more conspicuous than the radial polarization of the rings around them. This is for instance the case for the glory of plate 42. In the steam of a shower, it is easy to ascertain that the degree of polarization of the glory depends on its size: when the steam has just formed (so the droplets are small) the glory is large and its rings are considerably more polarized than when the glory has become smaller. In principle, it is even possible for *unpolarized* glories to occur sometimes. They may, moreover, differ from the usual glories by having dark, instead of bright, centres. It is not known if this type of glory often occurs in Nature.

The explanation of the glory is not simple: only since the 1970s has this phenomenon been fully understood. It has emerged that the light consists

Plates 43–45 *Top, left:* The diffraction corona around the sun. Only a part of this phenomenon is visible. The star-shaped appearance of the sun is caused by over-exposure in the camera. *Right:* Iridescent cloud exactly above the sun (§43). (Photograph P. P. Hattinga Verschure.) *Below, left:* The heiligenschein on bedewed reed, around the head of the photographer's shadow. (Photograph I. Können-Jongman.) None of these three phenomena is polarized.

mainly of two contributions. The first is a light-path with one internal reflection in the drop (like the one depicted in fig. 24 on p. 48), in which surface waves (§81) are responsible for the fact that the light rays can be scattered backwards. These surface waves cause a radial polarization of the rings of the glory. The second contribution is a *ten-fold* internal reflection of the light rays in the drops which is related to the formation of the tenth rainbow. Such light has a tangential polarization (§32). In small droplets the first contribution dominates; in larger ones the second contribution also becomes important. This explains the decrease of the radial polarization of the rings of the glory with increasing drop diameter. Interference of light causes a reversed direction of polarization, for both contributions, near the centre of the glory. The net result is a phenomenon with a very curious polarization. In this respect, the glory does not have an analogue in atmospheric optics.

Normally speaking, contributions like this (surface waves, light of the tenth rainbow) are very weak and hard to see in Nature. Since, however, the light near our shadow is amplified significantly, it does contribute to the glory. It is remarkable that in this way the tenth rainbow can actually be seen in Nature – albeit irrecognizably changed into the glory and manifesting itself merely as a decrease in the glory's polarization with increasing drop size.

43. The diffraction corona, the heiligenschein and iridescent clouds

To conclude this chapter three spectacles will be briefly described which do not show any polarization but should be included in a section on optical phenomena in the sky.

The unpolarized *diffraction corona* appears as a bright coloured aureole around a source of light (plate 43). It is much smaller than the halo (some degrees in diameter, just like the glory) and is very bright. It is best seen near the moon or near a terrestrial source of light; near the sun it is difficult to observe with the naked eye because of its brightness. Yet this is when it is at its most beautiful. It is usually noticed near the moon and it is almost the only optical phenomenon that maintains a wealth of colours by moonlight. One sees it appear when there are thin clouds in front of the moon. It is formed by diffraction by the cloud particles; its centre is vividly white, with a red edge round it. Round it appear coloured rings, which in favourable circumstances can sometimes recur but then do so with a continually decreasing intensity of light. The size of the rings is inversely proportional to the diameter of the particles, as in the case of the glory. When a second cloud passes in front of the moon, a change in the diameters of the rings can be noted; near the edge of a cloud the corona is often slightly egg-shaped (this also occurs with the glory). During a foggy night one often sees a corona appear around a lamp-post or around the head-lights of a motorcar. Coronas can also frequently be observed through blurred window panes; then, however, the centre is dark. In this case they are not formed by diffraction but rather by interference of light that has passed through several little 'holes' between the drops. Atmospheric coronas can be formed in both water-clouds and ice-clouds, but in the latter case the coloured rings are less well developed. Even around bright planets like Venus or Jupiter the corona can be seen; here, however, it is only the uncoloured centre that appears since the intensity of light in the coloured parts is too low for any observation. On the other hand, a complete diffraction corona has sometimes been seen around the totally eclipsed sun: then the solar corona was the source of light (see also §27).

The *heiligenschein* (plate 45) is seen as a white aureole around its own shadow; it can be best observed on dewy reeds or grass. This phenomenon is reminiscent of the glory, but without its colours and regular structure. It is mainly the result of the dew-drops acting as lenses, projecting a small image of the sun onto the blade behind it. Looking in the same direction as the rays of the sun, one perceives this bright spot in the lens itself amplified to the size of the drop. At some distance from one's shadow, however, the drop shows us only the dark part of the blade in the shadow of the drop. A second cause of the heiligenschein is that, when we watch along the same direction as the sun's rays, we will never see the shadows of grass-blades on other grass-blades (just as we never see shadows on the full moon; the full moon is after all nine times brighter than the half-moon!). Hence the heiligenschien may also appear on dry grass or even on rough surfaces, although it is considerably weaker than on bedewed grass. On the other hand, astronauts walking on the moon saw a bright heiligenschein: in this case glassy spheres in the arid lunar soil take over the role of the dew-drops. Like the diffraction corona, the heiligenschein is unpolarized.

Iridescent clouds commonly appear at 5–15° from the sun (plate 44). They are extremely bright, showing nacreous colours in irregular structures. This happens frequently in lenticularis clouds; dark sunglasses are necessary to aid observation. As with the diffraction corona, colours are formed by diffraction of sunlight by tiny particles of the clouds, the variety depending on the particle size and the angular distance from the sun. At 5–15° the colour is extremely sensitive to the particle size, so that small variations result in another colour being produced. Polarization, however, is absent in these clouds.

Natural scenery and objects around us

44. Introduction: a broad survey

Nature with her uplands and plains, her vegetation, her animals, and her objects, man made or not, displays a great variety of light effects and colours. Nearly all this light is caused by reflections on a multitude of objects around us. The colours we see are determined by the character of such objects, but the general appearance of Nature and the play of light and shadow rather depend upon the illumination of the moment. All the following provide an enormous scale of possibilities: sandy plains, woods, towns, grassland and mountains have very different characteristic appearances, which can change considerably when, for instance, the sun retreats leaving an overcast sky or when the sky clears after a shower and the sun shines upon a wet world.

Looking at the colours around us we find that in general greens (grass, growths) and yellows (sand) dominate the scenery; purples, pinks or oranges occur to a much smaller extent. This is particularly obvious in winter. In other seasons it is evident that flowers, butterflies, beetles and other insects above anything else display a chequered variety of colours and that any colour can be seen anywhere. Yet the overall image remains the same: some colours (e.g. green) predominate; others appear only, though strikingly, in certain isolated objects in the natural scenery.

The shades of polarization in the scenery bear some resemblance to the colour shades. Since most of the light we receive is reflected on all kinds of objects, there is also a predominant polarization, which is linear and tangentially directed with respect to the sun by unclouded skies. Smooth or polished objects have a white gloss, which is strongly polarized, but there is no gloss on rough surfaces and their polarization is fainter. This changes, however, after a shower: then, all objects are covered with a thin layer of water and they have a gloss as if they were polished. The polarization of their light increases considerably: the world after a shower is a great deal more strongly polarized than an arid plain. In this connection it is interesting that only one simple rule (Umov's: §§45 and 77) governs the degree of polarization of plains; in this chapter we shall often refer to this rule.

The polarization of objects, unlike their colour, depends mainly on the manner of illumination. Objects on which the sun shines directly have a tangential polarization; but under overcast skies or in the shade the polarization is in the main horizontally directed. At twilight, the source of light (the blue sky) is itself already strongly polarized, and this may give rise to different directions of polarization in the scenery around us, even occasionally producing some circular polarization. Yet the world can be considered to be tangentially or horizontally polarized; variations from this only occur with certain objects or by a special illumination. As flowers and butterflies multiply the riot of colour in Nature, so some isolated objects or certain special circumstances lead to a greater variety in the overall picture of polarization. For instance, flat, smooth objects such as the surface of still water may create reflections with unusual polarization: they function as mirrors and only the polarization of the light that happens to be reflected in the direction of the observer is important. Then the source of the light is by no means always the sun itself; it may just as well be a section of the blue polarized sky, and this may lead to varying polarizations in a predominantly tangentially polarized world. When sunlight is not reflected but is, say, refracted, the polarization will also differ from the usual tangential or horizontal polarization.

As mentioned above, the way in which polarization arises depends much more on the illumination than upon the nature of the material; almost all materials transform unpolarized light by reflection into the same kind of linearly polarized light, which is either horizontally or tangentially directed. Only the degree of polarization differs from material to material. But there is an exception to this: the reflected light off some beetles is almost totally circularly polarized, irrespective of the illumination! This is the sole event of its kind in Nature's polarization phenomena; a similar exception in the world of colour does not exist.

As a whole the polarization of the natural scenery and the objects in it is slightly fainter and the polarization pattern is more uniform than those of the sky and the exotic optical phenomena discussed in the preceding chapters. However, the polarization is very sensitive to lighting and humidity, and contrary to the case of optical phenomena, the overall situation can be understood by the application of a small number of simple rules.

45. Plains

Grass-, snow-, sand-plains and others have a tangential polarization in full sunlight; the polarization is strongest at about 90° from the sun (§ 72). Its maximum value, however, depends on the reflective capacity: a dark plain is more strongly polarized than a bright one. This is the so-called *Umov effect*, which applies to rough surfaces (§ 77). It is easy to see that snow, gypsum or white sand are much less polarized than, for example, the dark earth in a dry ditch, or asphalt. The effect can clearly be seen on different kinds of roads and I shall discuss this in the next section.

On some plains (e.g. grass) the separate particles can also have a white gloss, because they are still rather smooth. This gloss is often more intense than their own green colour and highly polarized. With a polarizing filter in a radial direction, this gloss will disappear almost entirely and only the warm colour of the plain itself will be visible. It can also be observed on the leaves of trees, which often have a stronger gloss than the blades of grass (§ 49 and plates 53–54 on p. 83).

The gloss and consequently the polarization are much stronger when a plain is wet with rain. Such plains glitter more (are whiter) than dry ones, and are, for just that reason, strongly polarized. Wet objects also have a deeper colour and are usually much darker, so that according to Umov's rule those parts which do not by chance sparkle in the direction of the observer, are polarized more strongly than normal! It is clearly illustrated by concrete, sand, mud, cobbles, gravel, rocky soil and many other examples. It is sometimes difficult to observe the polarization of bright objects when they are dry, but it is no problem at all when they are wet (plates 46–49).

Under overcast skies the tangential polarization of plains changes to horizontal, and generally the degree of polarization is somewhat reduced. In this case too Umov's rule is valid: dark plains have the greatest polarization. The only difference with unclouded skies is that the source of light is somewhat more diffuse and the 'average source of light' is exactly above us in the zenith. The polarization in these circumstances is also tangential with respect to the source, but now this corresponds to horizontal polarization.

At twilight the polarization of plains depends on the direction in which we are watching. At about 90° from the sun it is vertically directed, since the most strongly polarized section of the sky above it is the main light to be reflected, and is itself already vertically polarized. The contribution from this section of the sky to the reflection is greatest, although the light from other parts of the sky is also reflected by such a plain in the direction of the observer. The polarization in the direction of the set sun or exactly opposite to it, on the other hand, is horizontal: the light from the sky above it is less polarized, and hence the situation does not differ much from an overcast sky. However, the horizontal polarization of the plains will now be somewhat stronger compared to the latter situation, because the polarization of the blue sky above it in these directions of view is also mainly horizontal (fig. 19, p. 32). Somewhere between these directions, the plains show an unpolarized *neutral point*, which also occurs in the blue sky, in clouds or on an undulating water surface (see §§ 14, 23 and 54).

In the *shade*, during broad daylight, it is the blue sky that serves as a source of light, unless it is completely screened by objects. Shadowed plains usually show a horizontal polarization. Yet it need not always be the case, because light from the sky being polarized, though less strongly than at twilight, may induce changes. Since, however, the polarization direction of the light from the sky now forms a smaller angle with the horizon (§ 12), these alterations are not so large that they are immediately conspicuous. The contrast between the tangential polarization exactly outside the shade and the almost horizontal polarization in the shade (fig. 35) is often very striking.

In all these cases it is certainly worth studying in detail the direction and degree of polarization of plains in order to see how different they can be for various points of the compass, types of illumination and states of humidity and to find polarization effects other than those just described. An interesting example is to detect how the direction of polarization of a grassy plain can differ when lit not directly by the sun but indirectly via a reflection against a pane of glass. A polariscope is often useful for many kinds of observations, as weaker polarization can be easily detected with it. The observation of polarization, gloss and brightness and the use of Umov's rule deepens our insight into how light effects arise in our surroundings.

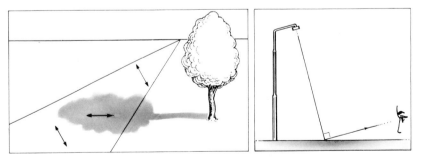

Fig. 35 A shadow has a direction of polarization different from that of a sunlit road.

Fig. 36 The glow on a dry road under a lamp-post is strongly polarized, because the light is reflected at about 90°.

46. Roads and mirages

A road is really an ideal example of a plain. All previously mentioned effects are clearly visible: the tangential polarization in sunlight, the horizontal polarization in shadows and by overcast skies, and the changing direction of polarization during twilight. The darker the road the greater is its polarization. New asphalt roads show the strongest polarization, concrete roads the lowest polarization; bricks lie more or less between the two. We can see the polarization increase clearly when the road is wet (plates 46–49); and marks on the road are often better noticed with polaroid sunglasses. There is a strong glitter and a high polarization when the road is covered by frost. The contrast between the horizontal and tangential polarizations inside and outside the shadows (§45) is sharply defined on wet roads, particularly when the shadow of an isolated object (e.g. a tree) is observed perpendicular to the sun's rays and the solar elevation is not too high (fig. 35).

By night the aspect of roads undergoes a change; the roads are then lit by many lamp-posts. Very wet roads reflect the sources of light as quiet puddles and these reflections may have a high degree of horizontal polarization. The reflections fade away when the road dries up, and a more or less diffuse glow remains just under the lamp-post. However, this glow is often strongly polarized: unless the lamp-post is too near, the patch under it has an angle of reflection of about 90° (fig. 36)! So on a wet road the reflection of a distant lamp-post is little polarized (grazing reflection) but a near reflection is, on the contrary, highly polarized; on a dry road the situation is reversed.

On hot days *mirages* may occur on roads. There seems to be a pool of water in the distance, which dissolves when we near it. Motorcars and other objects are mirrored in this 'pool'. Such mirages are also seen on plains and on the sea. They are, however, not real reflections but formed by refraction. The nearer to the ground, the warmer the air and the smaller its index of refraction: by this the direction of grazing rays of light is gradually changed to such an extent that they do not reach the ground but are deflected in an

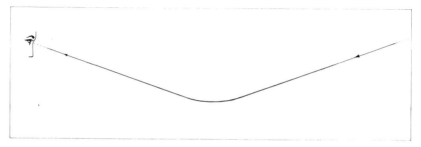

Fig. 37 Formation of a mirage above a hot plain.

Plates 46–49 *Top:* the dry world. *Below:* the wet world. The colours of the wet world are deeper and its polarization is stronger (§§45 and 46).

upward direction (fig. 37). This gives the same impression as a reflection. With such a gradual refraction (for that is indeed what it is) no polarization effects occur so that unpolarized light remains unpolarized. This makes a mirage different from a reflection on a real pool of water (§54). When polarized light is mirrored by a mirage, the degree and direction of polarization are not changed. This can be seen, when the mirage of a strongly polarized section of light from the sky appears.

47. Paper, frosted glass, houses, stones, mountains and metals

The reflected light from all these objects is linearly polarized, the degree of its polarization depending upon their smoothness, their wetness, how dark they are (the Umov rule), the angles of incidence and reflection of the light and naturally upon the manner of illumination. Again, in the case of diffuse lighting, as under an overcast sky, the direction of polarization in almost all cases is parallel to the horizon (see §§24 and 45). In the case of a sunlit object, the direction of polarization of the diffusly reflected light (to which the object owes its colour) is tangential with respect to the sun. If the object also has a gloss, this gloss is uncoloured and its polarization is horizontal with respect to the reflecting surface. For both the diffusly reflected light and the gloss, the degree of polarization is highest when the incident ray forms an angle of about 90° with the reflected ray.

If an object is rough enough, no gloss will appear on it. However, a number of rather rough objects do show a gloss, if the angle of observation is sufficiently oblique. This occurs with *newsprint, frosted glass, gravel* and many other fairly smooth objects. The more obliquely we look at a rough object, the rougher it must be to show *no* gloss! For blue light we must look more obliquely than for red light (a longer wavelength) in order to see the gloss. Hence, at a given angle, the reflection of an uncoloured lamp in such an object turns **red**. This can be seen best with frosted glass. When we look at a grazing angle at such glass, the reflection appears white; on looking straight at it, however, the sharp reflection and the gloss vanish and only a diffuse spot remains. The angle where the transition takes place and the red reflection appears depends on the ratio between the roughness of the surface and the wavelength of the light.

Besides frosted glass, newsprint is also appropriate for this kind of observation and for the study of the polarization of light reflected by rather rough surfaces. Looking at it obliquely, we find a horizontally polarized gloss; on looking more directly at it, the gloss goes. Meanwhile, its diffusly reflected light remains tangentially polarized with respect to the source of light, but the black ink has a considerably higher degree of polarization than the white blank parts of the newspaper. Moreover, the gloss near the printed parts will still be seen, when looked at more directly, while that near the blank parts will not: printer's ink has a smoother surface than the newsprint itself.

Similar observations can be taken with *wood*. Polished or wet wood may be so smooth that it always has a gloss which, of course, is horizontally polarized. By removing the gloss with a filter, we see once more the warm colour of the wood itself. Dry rough wood does not show such a gloss, and has a weaker polarization. Surfaces like that of fairly smooth board are intermediate types; looking obliquely at them, we observe a gloss, but looking straight at them we do not see any gloss at all. So, in the first case they seem smooth, in the second case they do not, although the material is the same in both cases. There are many objects around us which behave in this manner.

Bricks are rather rough and, when observed even at quite oblique angles, do not show any gloss if they are dry. The polarization of their diffusly reflected light is tangential with respect to the light source. So it does not matter if they are lying on the ground forming a road or upright forming a wall: a dry *house* has the same tangential polarization as a dry road. *Mountains* are also tangentially polarized, but high rocky mountains are so smooth that they can show a gloss with a high degree of polarization. The same is the case with *shingles, schist* and many boulders. But generally we see distant sunlit objects in a bluish haze, caused by the scattering of sunlight on the column of air between the observer and the object (see also §20). This haze is also tangentially polarized so that the visibility of these objects can be improved with a polarizing filter by extinguishing this foreground light (§§ 13, 16). Of course, the tangential polarization of the distant objects themselves is more or less veiled by this effect.

Finally, reflection on *metals* is also polarized, but less strongly than the gloss of other objects. The darker the metal, the stronger is its polarization (§ 76). However, maximum polarization occurs at a rather grazing reflection.

48. The motorcar

One can note many polarization effects on motorcars, as they are fairly smooth and glossy, consist partly of polished steel and often have doubly-refracting windscreens and (sometimes) rear windows. A great number of these polarization effects could also be seen on objects in the natural scenery, but the motorcar shows many of them on one single object, and is thus worthy of being discussed separately.

The direct reflection of the sun (visible as a clear, white, shining spot on the roof of the car for instance) is horizontally polarized; when the reflection comes from a polished metallic part of the car, its brightness is greater but the polarization is less (§ 76). Enamelled metal plates act as mirrors in which it is not always the sun but sometimes the sky or the street which is reflected. This reflected light is also horizontally polarized with respect to the reflecting plane but since, for example, the plane of the doors is vertical and that of the roof horizontal with respect to the road, their directions of polarization differ considerably from one another (fig. 38, see also § 51). As we rotate a polarizing filter, we can see other parts of the motorcar lighting up. Where the gloss is maximally extinguished, the colour of the enamel is deepest; when the gloss on the windscreens disappear, they become more transparent and we can easily see objects inside (plates 50–51). A polarizing filter also enables us to see colourful patterns appearing in the windscreens and rear-windows, these patterns are almost invisible without a filter. These colours are caused by the doubly-

Fig. 38 Directions of polarization on several planes of a motorcar.

Plates 50–51 Each of the several planes of a motorcar has another direction of polarization (§48).

refracting properties of the safety-glass, and can also be seen with a circularly polarizing filter. In §§52 and 53 I shall deal with observations which can be taken of this kind of material. Sunlit tyres show a tangential polarization, as may be expected from rough surfaces. This polarization is fairly strong, as predicted by the Umov rule for such dark objects.

When the blue polarized light from the sky reflects on the roof of a motorcar, the polarization direction of the reflected light is slightly rotated. Also, at a steep angle of incidence this plane is mirrored (§87, plate 52). Looking at twilight without a polarizing filter at the roof in the direction of the Brewster angle, we see a dark spot as a result of the extinction of the vertically polarized light from the sky by horizontal reflection (see also §15): there, for the most part, the gloss has disappeared. Thus, with the unaided eye, we can see that the light from the sky is polarized. Metal parts (bumpers, door-locks etc.) may have a *circular* polarization, particularly by twilight, since the incoming linearly polarized light from the sky can be converted by metal reflection into circularly polarized light (§91). Under favourable circumstances, it is very strong: the light of the chromium is largely extinguished with a proper filter, which one would not expect to happen so completely with a metal (see also plates 6–7 on p. 15).

49. Plants, animals and men

In the main, the polarization of the light of living things follows that of inanimate objects in Nature. Some *leaves*, for example, can be so smooth that, particularly at more oblique angles of incidence of light, they show a gloss which is horizontally polarized. In a bunch there can be many leaves oriented in many different directions: but the gloss of the bunch as a whole is tangentially polarized with respect to the sun (plates 53–54). Other leaves, flowers and vegetable material are often rougher, less glossy and have, consequently, less polarization; the same is true for smoother

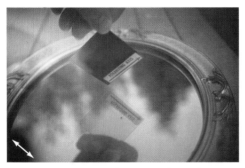

Plate 52 The plane of polarization mirrors at a steep angle of incidence (§§29 and 48). Here it has been demonstrated with an ordinary mirror where the effect occurs when the incident light is at a rather grazing angle (§91). A second polarizing filter is in front of the camera; its direction is indicated on the photograph.

Plates 53–54 A bunch of leaves has a tangentially polarized gloss, which can be extinguished with a polarizing filter (§49).

Plates 55–56. The rose chafer has a greenish metallic gloss which is left-handedly circularly polarized. L is the left-handed, R is the right-handed filter before the camera. The sparkling whiteness on the beetle remains visible without any change; it is linearly polarized (§50).

material upon which the sun is shining more perpendicularly. When there is no gloss, Umov's rule again holds: the darker the colour, the higher the polarization.

The abundant *hair* of many animals has, like the hair of man and the *feathers* of birds a strong gloss which is tangentially polarized. *Skin* has the same polarization as stones or leaves, the degree depending on its smoothness and its moistness (perspiration!). *Eyes* have a lustre that is strongly polarized.

All this suggests little that is new: living organisms generally polarize light in the same way as inanimate material does. It is interesting, however, that pieces of skin, bones, wings of insects, nails and other substances are doubly-refracting as a result of having a layered structure. So, when viewed between polarizing filters, they display colours (§93); the colours in nails, particularly, are clearly visible. What other biological materials have this chromatic polarization?

Some tropical *butterflies* have a magnificently coloured gloss, caused by many reflections on the wingscales. In this case an accidental combination of reflections can convert unpolarized light into circularly polarized light – which is, of course, only present on a minute spot on the wing and varies strongly with the angle of incidence of the light.

As far as polarized light is concerned, *insects* usually behave in the same way as other animals, hence sometimes showing a tangentially polarized gloss. The sole exception is dealt with below in §50.

Since there is virtually no difference between the polarization of inanimate objects and that of living things, it is also hard to ascertain by polarimetry whether there is (or was) any life in any place terrestrial (or extraterrestrial).

50. The circularly polarized gloss of certain beetles

The only beetles which have this unique property belong to the family of the Scarabaeidae, among them are the rose chafer (*Cetonia aurata*), the cock chafer (*Melolontha* sp.), the summer chafer (*Rhizotrogus solstitialis*), the garden chafer (*Phyllopertha horticola*) and some others. These beetles are partly or completely dark with a greenish or yellowish gloss, the gloss of the rose chafer being spread all over its body.

It turns out that the coloured gloss consists of *completely* left-handed circularly polarized light! With a left-handed filter, the gloss appears somewhat brighter; but when we observe such a beetle with a right-handed filter, the gloss has totally disappeared and the beetle is black (plates 55–56). This circular polarization occurs at any illumination, irrespective of whether or not the incident light is linearly polarized. However, when the incident light is linearly polarized and vibrating in an appropriate direction, the direct, white reflection against the body is absent and the circularly polarized gloss will unfold still more beautifully.

This gloss can be best observed on the rose chafer, which, contrary to what its latin name suggests, is really usually more greenish than yellow. It shows this circularly polarized gloss all over its body. In Western Europe, this insect can be found on strawberry plants, roses, peonies, lilacs, apple trees and on sandy soils from May to July. Its outward appearance is slightly reminiscent of the dung-beetle. Circular polarization can also be seen on the cock chafer, the summer chafer and the garden chafer: the head and other dark parts have a green gloss comprising left-handed light, but other parts, for example the brown wing cases, do not show this. Close scrutiny is needed to detect circularly polarized light on these beetles. The three species can be found from May to June – the garden chafer particularly in moorland and dunes, on the creeping willow and on all sorts of flowering plants; the cock chafer and the summer chafer especially on oaks.

Unfortunately, beetles like the rose chafer have become rarer in the course of the years. In former days it was so common that, because of its beautiful gloss, it was an object of trade for the young people of the flower market in Amsterdam. Nowadays it can only be found in the open country after a great deal of effort. Their wonderful circular polarization can, of course, also be studied in a museum, as the property is retained after the death of the specimen.

The cause of this circularly polarized light is briefly explained in §82. As far as is known, the above-mentioned insects are the only objects in Nature that convert unpolarized light directly into circularly polarized light; and because of this characteristic, many beetles of the Scarabaeidae can be readily distinguished from other families. The circular polarization is the more striking, since this form of polarization is so rare in Nature and the natural environment of these beetles does not usually show any trace of circular polarization. The biological function, if any, of this anomalous phenomenon is as yet unknown.

Finally, it seems that Nature is even able to create exceptions from this exception: sometimes a specimen reflects not left-handed but right-handed circularly polarized light! Such mutants, which need not stand out in any other way from their congenerics, are of course extremely rare.

51. Glass objects, panes, diamond, quartz and other transparent materials

The light effects created by these kinds of objects depend on the shape of the transmitting material. The simplest are those of a flat plane like a plate-glass window. If it is relatively dark behind a window, the light reflected by

Plates 57–58 By day, a plate-glass window can be made more transparent with a polarization filter; by suppressing the external reflection on the glass, the indoor objects become more visible. The darkest part of the right-hand picture corresponds to reflection at the Brewster angle (§§51 and 74).

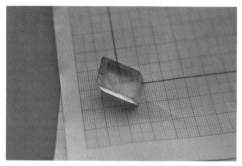

Plate 59 Everything is seen double through calcite ('drunken man's glass', §51).

the front side of the window is often so strong that we cannot look through the glass any more. This situation sometimes occurs when we try to look into a house or motorcar by day. The reflected light is horizontally polarized to a high degree so most of it can be extinguished with a polarizing filter: the pane becomes transparent and objects in the house become visible (plates 57–58 and 4–5 on p. 15). Polaroid sunglasses are equipped with polarizing filters which are vertically oriented with respect to the horizon in order to extinguish, for instance, the gloss of puddles on the road. Glass panes, however, are not lying on the surface of the earth but stand erect, so that their reflected light is vertically polarized with respect to the horizon. Polaroid sunglasses, therefore, not only do not eliminate the gloss but make looking into the house even more difficult – for the gloss to be extinguished the glasses must be rotated a quarter of a turn (fig. 39).

When we look out through a window from indoors (i.e. from dark to light), the pane hardly shows any reflections: the indoor objects are too sparingly lit. So the light coming from the outside can be seen without hindrance. As a result of the refractions of the pane, this light has now been polarized vertically with respect to the pane. When we look through the pane at a more grazing angle, the polarization becomes stronger. In the case of thick glass or double panes the incident light may be partially reflected several times between the refracting planes; this happens in such a way that near a small source of light (perhaps a remote lamp-post by night) extra images are seen next to the lamp: and so it appears double or multiple. But whereas the directly transmitted light shows vertical polarization, these extra images are horizontally polarized. By rotating a filter before one's eyes, the lamp and the extra images can be alternately extinguished (fig. 40).

With objects of more complex form there is a much larger range of possibilities in which light, via all kinds of refractions and reflections, can reach us. Total internal reflections may also occur in such objects and result in a greater brightness of the light; these light-paths appear as bright spots in the object. Other light-paths lead to sifting of colours which are projected as 'rainbows' onto the ceiling or which are visible as beautifully coloured sparkles on the object, as everybody knows who is the owner of a glass ash-tray or a chandelier.

The combination of refractions and reflections in such an object leads to a great variety of polarization effects. Total reflection (which yields the brightest spots) does not lead to polarization of the incident light but has the capacity to convert linearly polarized into circularly polarized light (§90). The greater the index of refraction, the greater the chance of such a conversion. Some 'light points' in a sunlit glass ash-tray show a strong circular polarization: the incident light has first been linearly polarized by non-total reflections and then converted into circularly polarized light by total reflections. This is clearly demonstrated by glass splinters but even more strongly by diamond or rutile (minerals with a very high index of refraction). Other sparkles in a transparent object tend to be unpolarized or linearly polarized. However, at twilight or in the shadow where the blue sky is the source of light, more sparkles are polarized, since under these conditions the incident light itself is already polarized. A diamond can show this circular polarization on several facets. The sense of rotation of

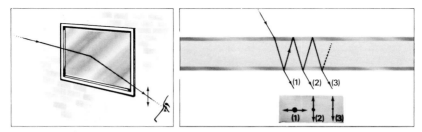

Fig. 39 Horizontally polarized light coming from an erect reflecting glass pane is vertically polarized with respect to the horizon.

Fig. 40 Light-passage through a thick glass pane. The extra images and the direct image are polarized in opposite direction.

the circular polarization varies from case to case and depends on the exact light-path followed by the rays through the material. Left-handed light is just as likely to be formed as right-handed light.

From an optical point of view, transparent objects can be divided roughly into two classes, *angular* and more or less *round* objects. A glass ornament in the shape of a donkey, for example, belongs to the second class; a glass ash-tray would fall into the first category. The optics of more or less round objects can be compared with those of a drop of rain, and the white light can be sorted into the colours of the rainbow. If the latter happens, the light-path must be: refraction→ reflection(s)→ refraction, and the glittering colours are linearly polarized (§27). The direction of polarization is tangential with respect to the sun. However, as the object is usually not completely bulb-shaped, it may be that the internal reflections are total; then there is virtually no polarization. Plastic bars, glass spheres or *objets d'art*, a glass of water, are examples of objects which can show the above-mentioned effects.

Angular objects, on the contrary, have the same optics as ice-crystals; here, as in the case of haloes, the colours can be sifted by two refractions without the necessity of reflections (§34). If the light-path still includes internal reflections, they will have often been total ones; so hardly any polarization occurs.

It is a different matter when the object is doubly-refracting: then the 'rainbows' on the wall are double and both of them are polarized with vibration directions perpendicular to each other. This can be seen with quartz-crystals. Putting a polarizing filter in the beam of light can cause one of the overlapping spots to disappear, and by rotating the filter the other will be extinguished (plate 36 on p. 63). Consequently, with a rotating polarizing filter, we can observe the coloured spot shifting to and fro. We came across the same phenomenon in atmospheric optics (the parhelia, which is a similar coloured spot, albeit projected in the sky; see §36). The double-refraction of quartz is, however, seven times stronger than that of ice and so the two polarized spots have much less tendency to overlap. A mineral like calcite is even more strongly doubly-refracting, and causes the components to be completely separated (plate 59). Such a mineral can project twice as many 'rainbows' on the ceiling as a glass object of the same shape; the pairs of attendant spots show a strong, always mutually opposing, polarization!

Finally, with a polarizing filter, colours can sometimes be seen in plastic objects as a result of chromatic polarization. This phenomenon can also be observed on doubly-refracting crystals like calcite. The colours are formed in the same way as those on the windscreens of motorcars, and will be treated in the next two sections.

52. Car windscreens and other doubly-refracting kinds of glass

Few motorists can never have seen the colourful 'spots' which appear in the windscreen, when they look through it wearing polaroid sunglasses (plate 60). We could infer that there is something strange about the windscreen: on ordinary window glass this kind of pattern never appears. The difference is that this windscreem glass is weakly doubly-refractive, which becomes evident in polarized light as colours: this is called *chromatic polarization* (§93).

For safety's sake internal mechanical stresses are created during the fabrication of windscreen glass for cars, by forcibly cooling the glass in some places for instance. In the case of an accident such glass will splinter into small chips like blocks, which are far less dangerous than big fragments. Because of these internal stresses the glass also becomes optically doubly-refracting (it has been locally 'stretched'), and as a result of this we can see the above-mentioned spots appearing in polarized light.

Colour phenomena like these come about only when the entering light is already polarized and when we are looking at the object with a polarizing filter (sunglasses). They happen when such a filter is put in front of the pane and another one behind it. In the case of the driver looking through his windscreen, the blue sky at about 90° from the sun or light reflections on the road act as sources of polarized light. The spots are then visible without a polarizing filter being put in front of the pane. They are noticeably absent, when one is driving more or less in the direction of the setting sun or under an overcast sky; in these cases light behind the screen is nearly unpolarized (see §24). The spots are clearly defined against dark roads and even more against wet roads, but they are almost invisible against a white concrete road or a sandy plain. Clouds are less polarized than the blue sky (see §20), and consequently when it is half-clouded, the patterns are proportionately less visible.

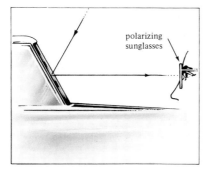

Fig. 41 Polarization of light increases by internal reflection, causing the coloured patterns in the windscreen of a motorcar to be more visible from outside than from within.

Plates 60–61 Motor-car windscreens often show coloured spots when one looks through them with a polarizing filter (*left*). As seen from outside, they are stronger (*right*). The colours are caused by double refraction of the windscreen (§52).

We can also see coloured spots on the windscreens of other cars, and they are often more striking than the patterns observed in our own windscreen (plate 61). This is not so surprising: here, part of the skylight has been reflected against the back of the glass so that the polarization has been considerably increased (fig. 41) and the spots thus become more pronounced. Apart from this internally reflected light, the same quantity of light is being reflected to us from the front side of the windscreen. This externally reflected light, however, results only in the ordinary gloss that appears on other parts of the car. This gloss can usually be almost completely extinguished with a polarizing filter, and then the windscreen merely shows the patterns caused by internal reflection, combined with chromatic polarization. With a polarizing filter before the eye in this position (vertical with respect to the windscreen so as to suppress the horizontally polarized gloss) the colour designs look most beautiful. For the rest, rotation of the filter causes a change of colours: on rotation of 90° they turn precisely complementary! This holds for all such observations of colour phenomena with a polarizing filter (§§65–69 and 92).

Linearly polarized light, passing through doubly-refractive matter is also partially converted into circularly polarized light (§93); the degree of conversion depends upon the colour of the light. So, we can also see with a circularly polarized filter that colourful spots appear on car windscreens. Left- and right-handed filters show complementary colours. The rest of the car, however, hardly changes at all, because ordinary gloss cannot be extinguished with a circular filter. This is in contrast to observations obtained by rotating a linear filter (§48).

On close inspection we shall find that these coloured patterns are also visible without the aid of a polarizing filter, when we look at the windscreen of a motorcar from the outside. This happens, when the blue polarized sky is the light source and we look at the pane obliquely: the internal reflection takes over the function of the filter (§15). Under an overcast sky the spots are also visible in the reflection of the windscreen in the bonnet; here, the internal reflection in the windscreen reflection is functioning as a polarizer and the reflection on the bonnet takes over the function of the filter. It is even possible to see the spots directly under an overcast sky, when we look at the pane very obliquely, because now the refraction of the light leaving the pane acts as a polarizing 'filter'. Of course, in the last case the spots are less obvious, because refraction is a less effective polarizer than reflection. All these cases are depicted in fig. 42; the place where the incident light is converted into linearly polarized light is indicated by 1, and the place where the light is passing along the second polarizer by 2.

Inside a motorcar these spots are also visible without a filter, when we look via a reflecting surface through the screen to a blue sky. Whereas the pane itself is transparent and clear to the naked eye, its reflection actually shows the spots. It does not seem real that an object and its reflection should have such divergent aspects (see also plate 67).

In addition to car windscreens, glass panes in trains and aeroplanes, some glass tables and strengthened window-glass (e.g. gymnasiums and on staircases) have also been made doubly-refractive, and so yield analogous observations. Moreover, glass that is mechanically subjected to stress, a glass tube which is vacuum pumped for example, also becomes doubly-refractive, and this can easily be observed with a polarized source of light and a polarizing filter. The double refraction is strongest in the places where the greatest stresses occur in the material; when the tension is released (by having the evacuated glass tube fill up with air), the double refraction also disappears. This property is utilized in industry to locate

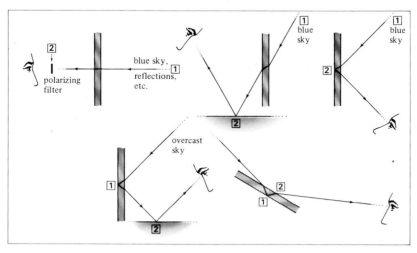

Fig. 42 Several situations under which coloured patterns become visible in car windscreens. At 1 the incident light becomes linearly polarized; where light passes the second polarizer has been indicated by 2.

weak places in materials. Besides glass, plastic and ice can also cause chromatic polarization (§53).

Sometimes 'safety-glass' is constructed by layering several sheets on top of each other. This kind of glass, which is also used for car windscreens has virtually no double refraction so that it is not possible to see any spots when wearing polaroid sunglasses (see e.g. plates 50–51). A type of safety-glass also exists in which certain parts have been extra-stressed like the part of the windscreen exactly opposite to the driver. In case of an accident this will be the first to shatter. The owner of the car can ascertain with a polarizing filter which kind of safety-glass has been used for his windscreen.

53. Flower-like frost patterns, ice-flakes and plastic: colour phenomena in polarized light. Demonstrations of slides with cellophane

Ice is a naturally doubly-refracting material, as we saw in the chapter on haloes. This double refraction causes parhelia to be shifted, when they are watched with a polarizing filter in various positions (§36). When this doubly-refractive material is put between *two* polarizing filters so that the incoming light is also polarized, chromatic polarization will take place and, consequently, colours will appear, exactly as on car windscreens. Minute *ice-flakes* from the refrigerator are a clear example. *Hailstones* can also be easily studied: because of the typical way in which they are formed, they show regular patterns which become visible, between polarizing filters, in many colours. These structures can be best seen when, by the warmth of the filters, the hailstones have melted to such an extent that only a flat slice is left.

For this kind of observation the blue sky may also function as a source of polarized light in which case the first filter is superfluous. Take a floating piece of ice out of a ditch and hold it against the sky or against a reflection of the sun in a lake: with a polarizing filter before the eye we can see all sorts of colours lighting up (plate 65).

Flower-like frost patterns on a window (nowadays rare because of central heating) can display chromatic polarization beautifully: with a polarizing filter we can see the white ice light up in unexpected colours, provided that there is a source of polarized light at the other side.

Plastic objects can also be doubly-refractive and, consequently, yield chromatic polarization, if they are internally stressed. This is nearly always the case with thick objects, which have not been cooled in a perfectly even way during the setting process. In polarized light the colours appear in places where the internal stresses are present. When the latter are uniformly distributed through the object, the colour patterns are similarly distributed (plate 66). Observations can be made with these objects similar to those described for car windscreens (see §52). One will notice, for example, that the reflection of the object against tiles or a glass pane shows colours, visible to the naked eye, which are absent on the object itself (plate 67). Small internal cracks, which are found particularly in older plastic objects, often show such colours to the naked eye since here the reflections on them take over the function of polarizing filters.

Thin small plastic bags are usually not doubly-refracting but they become so, when stretched with our hand. Between two polarizing filters we can observe that the places where the stretching took place, are lit up in fine colours. We see, as it were, the 'lines of stretch' appear.

Cellophane paper, as wrapped round boxes of chocolates, is also doubly-refracting, because of stretching during fabrication. The stretching has been done very evenly so that the double refraction is the same almost everywhere. Cellophane paper, folded two or three times, shows the resulting colours beautifully. It is also a very good example of what happens when such doubly-refracting material is made thicker – the colours do not follow the rules of mixture of paints (see also §93).

A slide with a polarizing filter in it on the same side as the lamp and a piece of cellophane folded against it on the side away from the lamp demonstrate splendidly what polarization is capable of. First, we see the colourless cellophane on the screen, but by holding a second polarizing filter before the lens in the beam of light, the same cellophane all at once displays magnificent colours: our 'polarization-blindness' is gone! Rotation of the filter causes a transition to other colour shades, until when the rotation is 90° everything has become complementary. Removal of the filter makes us polarization-blind again and then we see only the colourless cellophane (plates 62–64).

But there are still further possibilities for chromatic polarization: crystals can display a great deal more. The observations must be made under more controlled conditions, however, so they will be dealt with separately in §§65–70.

54. Water surfaces: puddles, lakes, the sea and the ocean

The aspect and the polarization of water surfaces depends not only on illumination but also to a high degree upon the presence or absence of waves. *Still water* acts as a large reflecting surface; the reflections of the sun and the sky are both horizontally polarized. Only the reflection of the most strongly polarized parts of the sky may have a slightly altered direction of polarization (see also §48). Consequently, the gloss of the water surface can for the most part be extinguished with a polarizing filter: the water becomes more transparent (see also plates 4–5 on p. 15). This is particularly clear for puddles on the road surface: the pavement under the puddle becomes much more discernible now. The linear polarization of the reflected light is

Plates 62–64 In polarized light cellophane displays colours which it does not possess without polarizing filters (§§53, 65, 92 and 93). *Below:* without filters. *Above:* between crossed filters (*left*) and parallel filters (*right*).

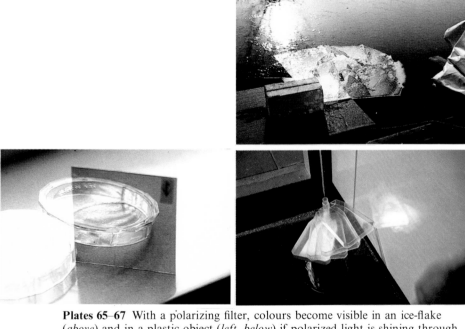

Plates 65–67 With a polarizing filter, colours become visible in an ice-flake (*above*) and in a plastic object (*left, below*) if polarized light is shining through them. (Photographs G. F. van Eijk.) *Right, below:* the reflection of plastic object displays colours which the object itself does not possess (§53).

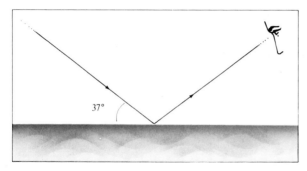

Fig. 43 Reflection on water at the Brewster angle, resulting in 100% polarized light.

100% for a reflection at the Brewster angle (§74), i.e. when we are looking at an angle of about 37° to the surface (fig. 43). The polarization is less, when we look at a more grazing or a steeper angle to the water surface (hence, farther away or nearer).

As mentioned several times before, the blue sky emits, at twilight and especially at about 90° from the sun, strongly polarized light with vibration direction vertical to the horizon. This light is less strongly reflected by water than the light coming from other sections of the sky, and is least when reflected at the Brewster angle. Here, without a filter, we see a 'dark spot' appear on the water surface (see §§15 and 48): to the left and to the right of this spot more light is reflected, because the polarization of the light from the sky is less there; above and below the spot, more light is reflected because the angle of incidence deviates considerably from the Brewster angle (plate 10 on p. 30).

Waves are formed, when the wind rises; the stronger the wind, the higher the waves. They are called wind-waves. When the wind falls, there are still waves on the sea for a considerable time; but the length of the waves grows larger and larger. These abating waves are called swell. The swell may have a direction different from that of the wind, which may again rise from another direction and create new wind-waves. On smaller areas of water surface, swell disappears quickly, but on oceans it may continue for days, becoming gradually lower and lower.

Undulating water looks quite different from still water: each spot on a wave reflects light from another part of the sky in our direction (fig. 44). But as the sun is by far the brightest part of the sky, its reflection completely dominates that of other parts of the sky. The polarization of an undulating water surface is, accordingly, tangentially directed with respect to the sun. Under an overcast sky the polarization is different because then the light is reflected from all sections of the sky. Yet, taken along the whole wave, the average polarization of the reflected light proves to be horizontal. Apparently, the light from directly above is more effectively reflected than

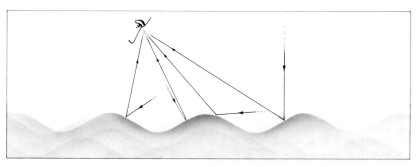

Fig. 44 Undulating water reflects light to the eye from all directions.

light from the sky coming from the left-hand or right-hand sides. Although in this case the polarization is never total, it is nevertheless rather strong. Finally, it is interesting that, at twilight, undulating water is horizontally polarized both in the direction of the sun and opposite to it, just as occurs under an overcast sky, but it is vertically polarized at 90° from the sun just like the blue sky itself! At about 40° from the sun, the horizontal polarization switches to vertical so that the gloss on the water is unpolarized. Thus the polarization has passed through a neutral point, as we also encountered in the blue sky (see §14). However, the 'negative polarization' (here, the horizontal polarization), is now much stronger than in blue sky, and the neutral point is consequently much farther away from the sun. We saw the same happen in the case of clouds (§23) and plains (§45).

On reaching the beach, waves change into surf. The white foam of the surf is almost unpolarized (remember the Umov rule). Sea-foam is also formed by gales in the middle of the sea and it can cover the sea completely. Consequently under these dismal circumstances, the light of the sea is also unpolarized.

55. The deep blue sea

The deep tropical oceans particularly display a beautiful blue colour; in fact the same blue light that is visible in the firmament. Particles of water scatter the sunlight in the same way as do particles of air in the atmosphere, so that the light from the ocean has the same properties as the light from the sky (see also §56) – it is blue and polarized. The blue light of the ocean is, therefore, sunlight, scattered under water by particles and shining upwards again out of the depth. Its polarization is maximal for light that has been scattered perpendicularly with respect to the incident beams; light, scattered in a forward or backward direction is unpolarized. Once again, the direction of polarization is tangential with respect to the sun, but it should be borne in mind that because of refraction of the sunlight at the water surface the sun's position seems higher when one is under water (see also §56).

This polarization can be best seen from a ship when the sea is still and when the position of the sun is half-way across the sky. Let us compare the sea in the direction of the sun with the sea exactly opposite to it (fig. 45). With the sun at our back the colour of the ocean water remains invariably bright blue, when a polarizing filter is rotated: we are looking along the direction of the sunbeams, and the scattered light out of the ocean is unpolarized. The only thing that changes is that the reflections upon the water can be extinguished. At the other side of the boat, hence in the direction of the sun, not only are the reflections extinguished but the sea takes on a deep indigo-blue colour, as the polarized scattered light out of the depths is now for the most part extinguished! This test alone is sufficient evidence that the blue of the oceans is not intrinsic but caused by the scattering of sunlight by the water molecules.

In still oceanic water, it is also possible to see a blue *heiligenschein* (§43), shining from the depths. It appears around our own shadow. The light of this little known phenomenon is not likely to be polarized.

56. The underwater world

The world of the fish differs completely from ours, as every diver knows. Under water everything seems much quieter; this world is characterized by a blue glow, particularly in deeper clean waters. The sun seems at a greater elevation when viewed from under water because the sunlight is refracted by the water surface. When the water is ruffled, the image of the sun dances up and down. The sunlight seen from under water is no longer unpolarized, but has become vertically polarized because of refraction, and this grows stronger the lower the position of the sun in the sky. The polarization maximum of about 30% occurs at sunrise and sunset. To the diver,

Fig. 45 Looking into the water in the direction of the sun, one perceives a polarized blue glow caused by scattering of sunlight upon the water particles at about 90°. Looking in the same direction as the sunbeams, one does not see this polarization.

however, the sun does not stand on the horizon at these moments but at an elevation of 41°, and displays a strong sifting of colour with red at the lower side and greenish blue at the upper side. Also the shape of the sun is much flattened, and the same happens to other objects that are very near to the real horizon. When we look lower than 41°, only the light of the slightly lit bottom is visible, being totally reflected against the water surface, and it is impossible to see objects above water.

As mentioned in the previous paragraph, the blue colour of the underwater world is caused by the scattering of sunlight by water molecules, in the same way as the blue of the sky is caused by scattering of light by molecules in the air. The polarization of this light shows, accordingly, the same pattern as that of the blue sky: the polarization direction is tangential with respect to the sun and the degree of polarization is maximal at 90° from the sun (see also §55). In quiet, clear water, the maximum degree of polarization amounts to about 60%. The polarization in turbulent water is consideraly lower and it drops readily from 60% to 30% or less. In these circumstances, although the quantity of light under water is greater, the whirls disturb the polarization. The polarization under water also decreases when the surface of the water is undulating, as the sunlight can be refracted in many directions by the waves. Then, there are no more parallel sunbeams underwater, hence the diminishing polarization of scattered light under the surface.

The similarity between the blue light from the sky and the blue scattered light under water is emphasized by the presence under water of areas with negative polarization, near the sun and the anti-solar point, and also of neutral points. Here too, the negative polarization is strongest in the area opposite to the sun and the degree of (negative) polarization can amount to about 15%.

Beneath an *overcast sky*, the direction and the degree of polarization of light change under water. Instead of the sun, we have a source of diffuse light which has an average direction of incidence perpendicular to the water. The direction of polarization under water now becomes horizontal with a maximum degree of 30% – a situation comparable to the polarization of the clouds themselves, under these circumstances (§24). At *great depths* (many tens of metres) the sun is almost invisible, even on a clear day. Then, the water is lit by light that has been scattered already by higher layers of waters, and therefore the illumination is more or less diffuse but with a preferential direction from above. The direction of polarization is once again horizontal, and the degree of polarization can amount to about 20%.

At *twilight* horizontal polarization of the light under water can be expected for the same reason. Now, however, the source of light is itself already linearly polarized, and when the sun is below the horizon, either in the west or the east, the direction of this polarization is north–south (§ 14). When we look under water in a northerly or southerly direction, the horizontal polarization of submarine scattered light corresponds to a west–east direction, exactly perpendicular to that of the light source: the incident light from the sky is, therefore, further weakened by the underwater scattering. This will not be the case in a westerly or easterly direction (§86). A diver should, consequently, be able to see through his mask that the scattered light is fainter at twilight in a northerly or southerly direction than it is to the west or east! As far as I know, this kind of observation (of intensities and polarization by twilight) has never been made; the more reason to try. Do altered directions of polarization also occur under these circumstances? Such determinations are not difficult, because they can just as easily be carried out in the shallows. The polarization of *objects* under water will hardly yield any new insights, but the fact that the blue light under water displays so many interesting possibilities for the observation of polarization must be an inducement to study the fascinating submarine world with a polarizing filter.

Besides linearly polarized light, circularly polarized light can also occur under water. The latter is formed indirectly by total reflection of the existing underwater linearly polarized light against the lower side of the water surface. This linearly polarized light may be the blue glow under water or the reflected light of objects. Such a total reflection does not display much circularly polarized light, even under the most favourable circumstances; the degree of circular polarization has a maximum of only 17%. It does not seem impossible, however, that with undulating water light is reflected *twice* against the lower side of a wave before reaching the observer. In this case the degree of polarization can be considerable and the circular character of the light easily observed. Under the most favourable circumstances (incident light being 100% linearly polarized; the angle of incidence of this light and its plane of vibration in the correct position with respect to the totally reflecting surfaces), the degree of circular polarization can amount to 62% after two total reflections! One should, however, keep in mind that under water such highly polarized circular light will only be visible at very special parts of such a wave. The chances of finding left-handed or right-handed circularly polarized light are equal (§90).

Nocturnal sources of light and artificial light

57. Introduction: the outward appearance of the world by night

When the sun has set so far that even twilight has disappeared, the world has undergone a real metamorphosis. In a deep, moonless night there is no longer any predominant source of light, but countless faint luminous points shed light upon us – the stars. Objects about us are hardly discernible. In principle, every point of light can only have its own characteristic polarization, but it appears that we must look hard for strongly polarized sources, for only a few objects in the firmament are radiating such light. Moreover, owing to the faintness of nocturnal sources of light, it is much more difficult to ascertain the polarization of such light than by bright daylight.

On a really dark night – which hardly ever occurs nowadays in our Western European civilization – we can see luminous phenomena appear in the nocturnal firmament which have no connection with the starlight itself: the aurora, the zodiacal light, the gegenschein and the airglow; and on clear summer nights there are also the noctilucent clouds (§25). It can easily be seen that some of these phenomena are, indeed, polarized; which adds to their beauty. Unfortunately, the moon or municipal illumination often impedes observation of the zodiacal light, the gegenschein and the airglow, while a vast auroral display is rare in Western Europe.

In addition to the celestial sources of light, faint terrestrial ones also appear sometimes on moonless nights. There is luminescence of the sea or of partially decayed wood, glow-worms in grass and (tropical) fire-flies for example; Will-o'-the-wisps sometimes appear near swamps. Flints struck against each other emit a blue spark. All these sources of light usually show virtually no polarization.

Our surroundings look quite different when the moon is above the horizon. Its cold light outshines that of the stars; objects about us are now quite clearly visible. The great contrast of a moonless night may be best experienced during a total lunar eclipse, when the predominent moonlight vanishes almost entirely within an hour. Although there is really no difference at all between the world by sunlight and the world by moonlight, anyone must find that the world by moonlight presents quite a different image. By day we take care not to look straight at the sun and so the illumination of the natural scenery and of the sky seems more characteristic. By night, however, the most striking object is the moon itself, and our eyes continually wander to it. The light of other objects is definitely secondary to it. What is more, when the light is faint, the human eye can hardly distinguish colours so that the nocturnal world by moonlight seems to us grey and white. All these factors cause the world to seem different to us by moonlight.

Since, however, the illumination by moonlight differs in no way from that by sunlight, the reflections about us and the observed phenomena of light in the sky are the same. But the 'blue' sky, the rainbow and other phenomena now seem colourless: only in the brightest optical phenomena, such as the diffraction corona and the paraselene, do we see colours. By now our gaze is also more attracted towards the bright phenomena which are near the source of light (the moon): the coloured diffraction corona round the moon is well-known; but hardly anyone, however, knows of the dim and colourless lunar rainbow. By day it is quite the reverse: nobody will look straight at the dazzling vicinity of the sun in order to detect a diffraction corona; but the colours of the rainbow are strikingly obvious, and known to everyone.

By moonlight, the polarization of the light round us underlines the essential similarity of illumination by day and by night. It will be easy, for example, to determine that, although the light from the sky seems milky white by moonlight, its polarization in no way differs from that of the blue sky by day (§ 18), that lunar rainbows are strongly polarized (§ 27) and that paraselenia shift, when observed through a polarizing filter (§ 36). In short, by moonlight we may expect exactly the same polarization phenomena of the firmament, the clouds, the optical phenomena and of the world round us as by sunlight, as depicted in previous chapters.

In our Western European civilization, it is usually of minor importance whether or not the moon is standing above the horizon: the artificial light of lamp-posts and of our houses outshines all other light. Artificial sources of light such as fluorescent light-tubes or TV-screens emit polarized light; and illuminated windows of houses, for example, show polarization as a result of refractions on glass panes (§§ 63 and 64). The reflection of lamplight on the pavement is also polarized, as discusssed in § 46.

On leaving a village or a town, we can still see the glow of city lights, particularly under an overcast sky. Here too polarization occurs, for the thousands of sources of light act together as one single source of light, and the light scattered against the clouds or the sky has the same characteristics as any other source of light, e.g. the sun or the moon.

Apart from the sun, few natural sources of light are so bright that they can outshine the city lights. The glow of a town, therefore, predominates at night; and only the light of a big blaze or a flash of lightning (or other sparks) is so intense that the city lights pale beside it. Fire and electrical discharges are not polarized, and polarization of the latter would hardly be noticed in any case because of the brevity of the discharge. Violent thunderstorms make us greatly aware of the influence of meteorological circumstances on the world's appearance: it would be difficult to conceive of a greater contrast with a bright moonless sky.

From this description we can see that the many sources of light in the nocturnal world offer the opportunity to make all kinds of observations, but that polarization of natural sources is seldom found. By night the world certainly has less polarization than by day, but by looking at all possibilities it becomes clear that some sources of light, or parts of them, may have a strong linear polarization. This striking property then differentiates such sources of light from others.

58. Stars, planets and other celestial bodies

There is no real difference between an average *star* and the *sun*: both are gigantic glowing gas spheres in infinite space, emitting a huge quantity of light. As our distance from the stars is so great, we merely see them as tiny luminous points, which by day are outshone by the light of our nearest star, the sun. The sun emits unpolarized light (§ 79), and so also do the stars. On its long way to us, however, starlight meets many particles of dust, which, under the influence of magnetic fields, have a preferential orientation in space. As a result, the starlight reaches the Earth more or less polarized, but this polarization is so low (at most a few percent) that it cannot be observed with simple devices (§ 84).

Unlike the sun and the stars, the moon, planets and comets do not themselves emit light but reflect sunlight to us. In these cases, linear polarization is very likely especially when the reflection occurs at about an angle of 90°, the celestial body being then in its first or last quarter. However, only the moon, Mercury, Venus, the comets and artificial satellites are capable of this. The *outer planets* (Mars, Jupiter, Saturn etc.) always appear to us to be almost 'full', and for this reason show hardly any polarization – again only a few percent or less, and therefore not visible with simple devices.

The surface of the *moon* consists of rough stones and dust. Its polarization is not nearly so strong as that expected from a smooth reflecting surface. The maximum degree of polarization occurs when the moon is slightly more than half-lit: averaged over the whole lunar disc, however, it is no more than about 10%. The direction of polarization is tangential with respect to the sun, hence directed 'from above downwards'. It is difficult to see the polarization, because the moon is such a small object. Dark areas are more polarized than light ones (the Umov rule) and the degree of polarization may amount to 20% or more for the maria, which is comparable to that of terrestrial sand-plains. Indeed, the polarization is visible with careful scrutiny.

The *Earth-shine* of the moon (the illumination of the dark section of the moon) also proves to be faintly polarized. Here as well, the polarization is maximal in the first and last quarters, when the light is rather dim. The maximum degree of polarization is also about 10%. The Earth-shine comes from the Earth with its oceans, clouds and plains. The polarization of the light of the planet *Earth* as a whole can be considerable (about 40% maximum), but after reflection against the dust-covered moon this light loses a large part of its polarization (§89).

The *full moon* is unpolarized: the sunbeams are vertically reflected to us by the lunar surface, and this situation is not changed during a *lunar eclipse*.

In favourable circumstances, *comets* can have a degree of polarization of 30% or more. Most of the light of a comet comes from its tail, and here, too, it is really sunlight that is scattered to us by gas molecules. Some *artificial satellites* can also be strongly polarized, to 40% or more. It is, however, not so easy to observe the polarization of these rather faint, rapidly moving objects. The degree of polarization of artificial satellites depends on the materials of which they are constructed, and there are many satellites that have hardly any polarization.

The polarization of *Mercury* is about the same as that of the moon: maximum about 10%. The structures of their surfaces are comparable. However, it is even harder to observe the polarization of Mercury, with its much fainter light, than that of the moon, indeed almost impossible for amateurs. *Venus* with its dense atmosphere is virtually unpolarized: in fact, its degree of polarization is even lower than that of Mercury.

Apart from what has already been mentioned, there is little else in the nocturnal sky which shows any polarization. Some gas clouds in space light up, because they scatter light coming from a nearby star, and this light is polarized. Yet the degree of polarization is rather low (usually about 20%) and extremely difficult to observe because of the faintness of their light. An example of such a cloud is VY CMa in the constellation of the Greater Dog.

An interesting challenge, however, is the polarization of the *Crab Nebula* M1 in Taurus. It is a remnant of a supernova explosion which took place in 1054; the then exploded star is now the famous pulsar which is in the nebula. The light of the Crab Nebula is synchroton radiation (§98), which is locally highly polarized (to about 70%!). The direction of polarization depends upon that of the local magnetic fields, and it can vary greatly from place to place. The strongest polarization is found at the edges of the nebula, where, however, the light intensity is less than at the centre. The shape of the nebula changes when we watch it through a rotating polarizing filter (plates 88–89 on p. 126). A big telescope is needed for observing this effect, and the polarizing filter must be put in front of the mirror because otherwise the polarization may be changed by reflections in the optical equipment. The observation is not easy to make but worth while because of the way in which the polarization is displayed. Together with the solar corona (§19) and comets it is the only astronomical object with a really strong polarization.

59. Aurora (plate 68)

In many parts of Europe, the aurora is unfortunately a rare phenomenon, but in Scandinavia, North America, and also at the South Pole it is frequently visible. It comprises luminescent bands, arcs and streamers, which sometimes have a typical curtain-like structure and periodically light up in a rhythmic fashion. The light comes from very great heights in the atmosphere (100 to 300 km). The light of an aurora is often not very intense (like that of the Milky Way), but sometimes it becomes very bright, and its beautiful colours are then visible.

The colour that the phenomenon takes on bears a close relationship to the height from which the light descends. The highest part of the aurora is red, the next part is green and the lowest part (at about 95 km) usually emits purplish light. On the transition from green to red, the light sometimes looks orange or yellow as a result of the mixture of the green and red. The light comes from molecules and atoms which have been excited by fast-moving electrons and atoms from the sun; during their decay they emit a characteristic light. The purple light originates from N_2 and N_2^+ molecules, which decay rapidly after excitation (10^{-8} s). In the higher layers of the atmosphere we see mainly green light from oxygen atoms; this transition is termed 'forbidden' and the atoms decay only after an average of 0.7 s. Only where the atmosphere is so rarefied that these atoms can decay before colliding with other particles, does this light become predominant. The long lifetime of the excited oxygen atoms can sometimes be seen in a pulsating aurora: aurora beams, speeding across the sky, have a purple front (from molecular light) and a green tail. At greater heights, an atom can fly about longer before colliding; here we can see the red light which originates from another even more strongly forbidden transition in oxygen atoms, where the excited atoms have a mean life-time of as much as 2 min. At greater heights in the atmosphere this light is so bright that it dominates the other transitions completely.

The red higher part of the aurora is polarized, the lower green and purple parts of it are not. The direction of polarization is perpendicular to that of the magnetic field of the Earth, hence directed mainly west–east. The polarization is maximal when we look perpendicularly to the magnetic zenith, which is at a height of about 70° in the northern sky for central England. Theoretically, the degree of polarization of the red light can amount to as much as 60%. The best chance to see this polarization in Western Europe is low near the northern horizon where the degree of polarization must be about 55%. In the case of rare bright displays, which may even extend in Western Europe to the south of the firmament, the most favourable point is in the south at a height of roughly 20°.

This polarization is related to the Zeeman effect, and manifests itself as a consequence of the magnetic field of the Earth (§97). However, not every atomic transition is sensitive to it, and that is why only the red aurora has a considerable degree of polarization. The aurora is the only phenomenon of this kind that shows polarization in Nature; other electrical effects do not show it to any noticeable degree. The reason is that other discharges (lightning for example) take place at a much higher atmospheric pressure so that the influence of magnetic fields is negligible.

It seems probable that *shifting* of colour in the aurora can be seen with a polarizing filter: when the filter is held in such a way that the red has been maximally extinguished, the (unpolarized) green must become relatively brighter. The yellow part of the aurora (where red and green are mixed) will consequently have to shift up or down, according to the position of the filter, since the ratio between red and green light has changed. So, for a minimal transmission of the filter, the lower part of the red will have to turn yellowish and the yellow to greenish: the effect will be reverse in the case of a maximum transmission.

60. Zodiacal light, gegenschein and airglow

Even more so than the aurora, these phenomena are visible only on deep, moonless nights at great distances from city lights. The *zodiacal light* is the brightest among the three and is outlined against the sky as a luminous wedge-shaped triangle in the position where the sun is below the horizon (plate 69). This light is brightest along the ecliptic,[1] which is also the direction in which the triangle points. This means that the phenomenon is best seen, when the ecliptic has a steep angle to the horizon, i.e. in spring after sunset in the west, and in autumn before sunrise in the east. The most favourable circumstances for observation are those with the position of the sun about 20–30° below the horizon; this means about three hours before sunrise or after sunset. Aeroplanes sometimes offer a magnificent view of the zodiacal light, whose brightness may be comparable to that of the Milky Way.

[1] The ecliptic is the line across the sky along which the sun moves in a year.

Plate 68 Auroral display, seen from the Antarctic. The low green aurora on this picture is unpolarized; the higher red one (not visible here) is linearly ▶ polarized (§59). (Photograph C. W. van Vliet.)

Plate 69 The zodiacal light as seen from some height above the moon by space vehicle Apollo 15. Its light is polarized (§60). (Photograph National Space Data Centre A.)

The zodiacal light has extraterrestrial origins and is also visible from the moon (plate 69). It comes into being by the scattering of sunlight on countless minute dust particles in the solar system which are concentrated in a huge disc-shaped area around the sun. The disc lies in the plane of the orbits of the planets so that we see its scattered light extending along the ecliptic.

The zodiacal light is, as may be expected from scattered light, polarized – tangentially with respect to the sun. Thus it differs from the unpolarized light of the Milky Way, which is produced directly by millions of faint stars. The degree of polarization of the zodiacal light amounts to a maximum of about 15–20%. These values can be found at about 60° from the sun, but the polarization is hardly less closer to the sun or farther away from it. At sea level, the zodiacal light is usually so faint that its polarization is hardly discernible. But from a high mountain this light can sometimes be surprisingly bright and then its polarization is distinctly visible, as I know from experience.

On very dark winter nights one can see that the zodiacal light extends like a thin light-bridge all along the ecliptic, while continually decreasing in intensity of light and in polarization. Near the point exactly opposite to the sun, the brightness slightly increases again: this is the *gegenschein* (literally opposition light). This gleam is analogous to the heiligenschein (§ 43): an increase of the brightness near the observer's shadow. The gegenschein is usually a very faint and difficult observable phenomenon. But again, from a high mountain, it can sometimes be relatively bright and is easily found as an area of slightly enhanced luminosity of several degrees in diameter, outlined against the dark sky. Its light is unpolarized, just like that of the heiligenschein.

Finally, there is also light in the nocturnal sky which does not come from the stars but from the highest layers of the atmosphere. It is called the *airglow*, and in favourable circumstances it can be rather bright. It is unpolarized.

61. The phosphorescent sea, wood, fireflies etc.

On moonless nights, when the sea is warm, particularly in autumn, the water can be seen to emit light of a greenish colour. It does not come from the seawater itself, but from minuscule animals that light up when oxygen mixes with the water. They light up for a very short time but, because there are such large quantities of them, the radiance of the sea looks constant. Near the breakers there is an optimum mix of air and water, and the light is at its most intense. But on the open sea, without any surf, we can also see luminescent spots floating through the water. If we rub our feet through the wet sand on the edge of a luminescent sea, we will see the minute animals light up separately like sparks.

Their light itself is unpolarized, but it is refracted by the seawater surface before reaching us and thus becomes vertically polarized (§ 79). Near the breakers, however, the surface of the sea is not smooth and polarization is absent but this light is polarized on the calmer open sea. This polarization is strongest when one looks at the water at a grazing angle, and theoretically it can amount to about 30 %. However, it is not easy to see the polarization of the phosphorescent sea because of the faintness of its light.

In a humid atmosphere, potatoes, freshly killed fish, decaying wood and leaves can also emit a faint light, the colour of which is also greenish. As with the sea, this is normally caused by the chemiluminescence of minute animals or bacteria. On wood, however, it is the hyphae of the honey mushroom that are the real source of light: an owl, crawling out of a decaying hollow tree, has these hyphae on its wings and also emits light, which is a ghostly sight. In the tropics fireflies fly about as small points of light, and in Europe there are fluorescent glow-worms. The light of all these plants and animals is itself unpolarized and because here it is not refracted by a smooth surface polarization remains absent.

62. Incandescent objects, fire, sparks and lightning

Fire, sparks and *lightning* (plate 75) can be considered to be the only nocturnal sources of light, which, though sometimes only for a short moment, can really light up their surroundings at night. All these sources of light are unpolarized, just like sparks from *flints* and *St Elmo's fire*, which is seen as a flame at the top of protruding objects in thundery weather (plate 76).

The situation is different with *incandescent metals*: when the surface is clean their glow is strongly polarized with its polarization direction vertical with respect to the surface. This is not generally known. When we look perpendicularly at the surface, the glow is unpolarized, but the more grazing the angle of observation, the stronger is the polarization (§80). The degree of polarization can even amount to a maximum of 90%! This polarization arises, however, only if the metal surface remains clean, even when it is red-hot. This applies to precious metals like silver or platinum where it is easy to see the polarization. The surface of a *streaming* liquid base metal, like steel or iron also remains clean and the glow is strongly polarized; as such a liquid may have small waves, the polarization may differ from place to place on the surface (plates 70–71). Another method used to prevent surface impurities on base metal is placing the metal in a vacuum (§64). But polarization is absent, when such precautions are not taken, because the surface is covered by layers of oxide. This means that the glow of a heated silver sheet appears strongly polarized, whereas the glow of the steel pair of tweezers holding the sheet does not show any trace of

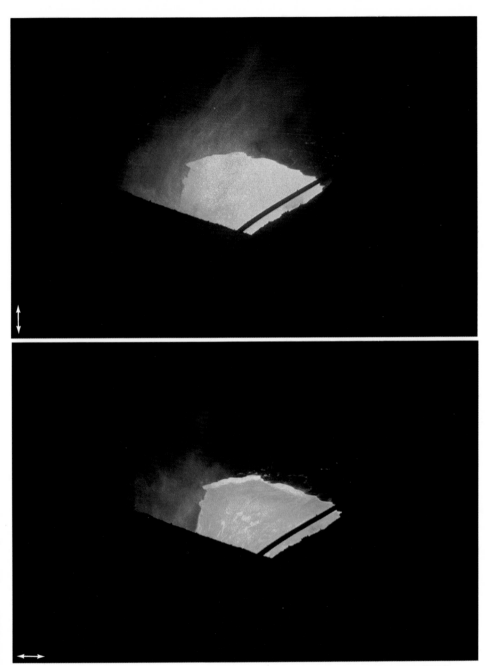

Plates 70–71 Streaming incandescent iron. The glow is vertically polarized (§62). With a horizontally directed polarizing filter (*below*) small waves become more visible. The side of the glowing stream moves slowly and oxide can be formed on it, so that polarization is absent there. The difference in the shape of the cloud above the iron is not caused by polarization, but resulted from the time lag between the pictures.

polarization. It is an almost unearthly state when one can nearly completely extinguish the glow of a heated sheet of silver with a polarizing filter, while the silver still maintains the yellowish colour resulting from its high temperature.

The polarization arises because the glow does not originate from the surface itself but from a (minimal) depth inside the metal. On leaving the surface, the light is refracted and becomes vertically polarized. The degree of polarization is, however, much higher than that resulting from refraction on glass for instance: incandescent glass also has a vertically polarized glow caused by refraction, but here the degree of polarization can never be higher than 40%.

The glow of streaming *lava* or hot stones during a volcanic eruption may be expected to have some vertical polarization, provided that the surface of the glowing matter is smooth. It is doubtful, however, if anyone would be sufficiently self-possessed to take such observations under these circumstances!

It turns out that a heated dichroic crystal like tourmaline also emits a polarized glow, even when we look straight at its surface. Then the polarization is perpendicular to the polarization which appears with transmitted light (§84). The stronger the dichroism, the higher the degree of polarization of the emitted light; it is about 40% for tourmaline.

63. Sources of light in a town

It is worth while to have a polarizing filter when walking through an illuminated town, because there is so much polarized light about us. The following is a brief description of the most important effects. Of course, many of the phenomena caused by artificial light are basically the same as those caused by sunlight, which have been dealt with in previous chapters.

Illuminated *windows* emit polarized light. The polarization direction is vertical with respect to the surface of the glass and consequently, in general, parallel to the ground. This polarization is caused by two refractions on the window pane and consequently can be rather strong (§79). The more obliquely we look at the panes, the stronger is the polarization. There is, of course, no polarization when we look straight through the pane. *Street-lamps* can also emit polarized light when we look at them obliquely, provided that their glass or plastic casings are smooth. The *reflection* of light from the street-lamps against the pavement or the houses is also polarized; the reflecting surfaces are rough, and consequently, the polarization direction is tangentially directed with respect to the lamp. (See also §46.) Directly under the street-lamps the brightness of the reflected light is greatest. Here, the tangential direction of its polarization corresponds to parallel with the ground, but away from this spot the plane of polarization makes an angle with the road. So, if the road is lit by many lamp-posts, the direction of polarization may differ from place to place. When a glaring beam of light from a search-light or a projector pierces the dark, this too is tangentially polarized: the beam is of course only visible because the light is scattered by the particles in the air. On leaving a town, we can still see the *glow* of city lights which, like all scattered light, is tangentially polarized; this is particularly visible in clear weather.

Artificial illumination at night has, in principle, the same phenomena of polarization of reflected or scattered light as can be seen in bright sunlight. The only real difference is that the source of light is either much nearer or consists of a great number of individual smaller light sources.

64. Artificial sources of light, such as lamps, dial-plates of watches, fluorescent light-tubes and TV-tubes

The light of *fluorescent light-tubes, TV-tubes, lamps* and other artificial sources of light is linearly polarized, with its direction vertical with respect to the surface of their outer covering of glass or plastic through which they shine. So the reason for the polarization is not that the light itself in these objects is already polarized (the discharge in a fluorescent light-tube, a sodium-vapour lamp or a mercury-vapour lamp is totally unpolarized), but that this light becomes polarized as a result of the refractions on the glass or plastic around it. As there are two refractions involved, the polarization can be considerable and may amount to a maximum of 60% for glass (§ 79). This is only the case, when one looks at the glass very obliquely and the light is, consequently, maximally refracted. More often we look at the glass in a considerably more direct manner and so the light is much less polarized. Moreover, since the shape of such a source is usually somewhat circular, the light of these sources *as a whole* is virtually unpolarized. However, it is easy to see where strong polarization is present within a source of light. With a polarizing filter, a dark line appears on a fluorescent light-tube, and shifts along the tube when the filter is rotated (plates 72–73). By looking obliquely at TV-screens we can observe their polarization. The edges of *incandescent bulbs* have a clear polarization: the places where the light is maximally extinguished with a filter co-rotate when the filter is rotated.

Some incandescent bulbs have been made not of frosted glass but of transparent glass. In this case the filament is visible and when heated to a high temperature by an electric current it glows red or yellow. If this filament, which is usually made of tungsten, is not too thin, the polarization of the metal itself may be observed, as described in §62. Tungsten is not a precious metal, but in vacuum its surface may remain clean. Its light is then vertically polarized with respect to its surface, and the polarization can be

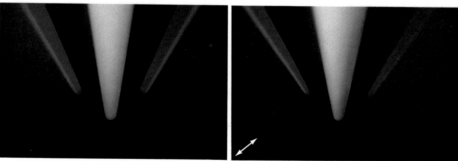

Plates 72–73 Fluorescent lighting is polarized. *Left* without and *right* with polarizing filter. Then, a dark band appears over the tube (§64).

Plate 74 A circular filter does not transmit its own reflected light, but does transmit light which has come from a filter of opposite handedness next to it (§§88 and 91). Such circular filters can be placed before sources of light to extinguish reflections (§64).

Plate 75 Lightning belongs to the few nocturnal sources of light, capable of outshining city lights. It has no polarization (§62).

Plate 76 St Elmo's fire appears here in thundery weather as a flame on an outcrop. It is unpolarized (§62). (Photograph G. Doeksen.)

very strong. As, however, a filament usually has an intricate helical structure, it is less simple to observe than might be supposed. In many cases, however, an incandescent bulb may be filled with gas, and then the metal surface does not remain clean so that polarization is absent.

The red shining *numerals* of an electronic counter often prove to be linearly polarized. The polarization is caused by a circular polarizing filter that is put in front of the numerals in order to suppress inconvenient light reflections against the dial-plate (plate 74). Such a filter consists of a combination of a linear filter and a sheet that converts linearly polarized into circularly polarized light. As this sheet has been fitted at the side of the source of light, we are looking at the linear filter and therefore we see polarized light (§§ 7 and 88). Without the filter, the red light of the numerals would be unpolarized.

The green shining dial-plate of an ordinary watch emits unpolarized light; but some polarization arises by refractions in the watch-glass, as is the case with lamps. On the other hand, the numerals of a modern digital watch are strongly polarized. Here, it is essential that a polarizer is placed in front of the watch in order to make the numerals visible, since they consist of liquid crystals which rotate the plane of polarization (§95). An electric field controls this rotation and determines which number should be displayed. If one rotates the polarizing filter in front of the numerals a quarter of a turn, the complementary view becomes visible: white numerals against a black background!

Colour phenomena of minerals in polarized light

65. Introduction

Doubly-refractive and optically active materials (§92) show colour phenomena when the entering light is polarized and when the emerging light is
observed with a polarizing filter. If one of these conditions is not fulfilled,
such material often seems glossily transparent and colourless, and so the
outer appearance undergoes a marked change, when the circumstances are
appropriate (see also plates 62–64 on p. 94). We see these colours appear
on car windscreens and in flower-like frost patterns (§§52 and 53), both of
which are doubly-refractive and, consequently, show chromatic polarization (§93). The effects are still more striking when a flat piece of mineral is
kept between two polarizing filters: because of its regular crystal structure,
the colours are now arranged in beautiful regular patterns! There are
hardly any minerals that do not show such interference patterns, since
almost all crystal classes (with the exception of the cubic one) are doubly-
refracting.

In contrast to the polarization effects that were discussed in previous
chapters, we have to do more than just look through a filter in order to
observe colour effects in minerals: a thin sheet of the mineral must be
obtained and the conditions mentioned at the beginning of this section
must be fulfilled. This kind of observation, then, comes more under the
heading of simple experiments than straightforward polarization effects in
Nature. The pains one has to take are amply repaid by the magnificent
regular patterns and colour effects which appear, examples of which have
been given in plates 77–85, and which are the subject of this chapter.

The patterns which appear in minerals in polarized light depend on the
type of crystal. The patterns of uniaxial minerals (§83) are different from
those of biaxial minerals, and any optical activity in the material will
influence the shape of the pattern (§95). Moreover, there are some minerals

which are pleochroic (dichroic or trichroic, §84), and they give rise to a quite different class of phenomena. Exactly how the patterns appear also depends on the filters between which the mineral is held; it is easiest to distinguish between the combinations linear–linear, linear–circular and circular–circular. The patterns with crossed and parallel filters are, however, always complementary, which means that black is replaced by white and, for example, red by green, while the shape of the figure remains unaltered.

Few know how easily these patterns can be seen: in most physics texts, the chapter on interference effects in convergent polarized light contains a sketch of a rather intricate frame with which observations can be taken and, in addition, a picture of a regular pattern taken in monochromatic light. It is usually not mentioned that these same patterns can also be observed by simply keeping a flat piece of mineral between polarizing filters close to the naked eye and that these patterns are beautifully coloured when white light is used instead of monochromatic! They are most striking when one looks more or less in the direction of an optical axis (§§83 and 94), but it is also worth looking in other directions. Some minerals will show patterns clearly when we look through the intact crystal, but for others flat sheets of the mineral are needed.

Though each mineral displays its own specific patterns, yet overall there is a certain system: all uniaxial minerals, for example, form similar patterns. In this chapter the patterns which can be expected will be described briefly for some minerals which are representative of a certain category of phenomena, without going into details concerning the explanation of the patterns (see §§92–95). The exact explanation of some effects is rather complicated and does not fit in the scheme of this book, but one may become fascinated by the phenomena shown by minerals in polarized light without any detailed knowledge. Finally, it is worth while experimenting, for example, pile up sheets of different minerals and observe what happens to the interference patterns: you will see surprising effects, not described in textbooks, particularly when one of the minerals is optically active (like quartz).

66. How to make observations

We need a sheet of mineral with a thickness of a few millimetres. The most beautiful patterns will appear when the sheet is placed so that we can look in the direction of the optical axis (axes). For quartz crystals it means that the sheet must have been cut perpendicularly to the six-sided face. If the crystal is to remain unimpaired the pyramidal ends may often handicap the observation. But this difficulty can be overcome by putting the crystal into a glass of water: the refraction of the light on the faces of the crystal is now

Plates 77–79 Interference patterns in calcite. *Top left:* between crossed linear filters. *Top right:* between parallel linear filters. *Below:* between one linear and one circular filter (§67).

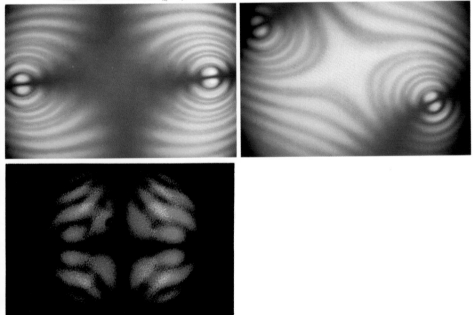

Plates 80–82 *Top:* aragonite between crossed linear filters, in two positions. *Below:* brookite between crossed linear filters shows a remarkable pattern (§68).

reduced. The patterns remain more or less intact, as their shapes are not so strongly dependent on the length of the crystal.

The regular patterns become visible when the sheet is held between polarizing filters and close to the eye. We can omit the foremost filter, when we look directly at a polarized light source, such as a reflection or the blue sky. Some crystals are twinned, which means that they consist of two or more individuals which have been grown together. In this case, patterns can sometimes be seen without filters: reflections on internal faces between the individuals in such crystals take over the function of the polarizing filters.

67. Uniaxial minerals, e.g. calcite

When we look at a uniaxial crystal between two linearly polarizing filters, concentric coloured circles appear in the direction of the optical axis. These circles are intersected by two bars, forming a cross; the bars are called *isogyres*. With crossed filters they are black; with parallel filters they are white (fig. 46 and plates 77–78). One isogyre is in the direction of the plane of vibration of the entering light; the other is perpendicular to it. Because it produces isogyres, the polarized light demonstrates that it does not vibrate in all directions and is, consequently, anisotropic (§5). There is no double-refraction in the exact direction of the optical axis, and the polarizing filters transmit light or not, as they would do without the mineral being put between them.

The isogyres will fade when one of the filters is replaced by a circular one. We now see that the circles are deformed: in pairs of adjacent quadrants bounded by the isogyres, the circles have different radii. In comparison with the original situation, the circles, e.g. in the first and third quadrants (the 'northeast' and 'southwest' ones), have smaller radii, and in the second and fourth quadrants (the 'northwest' and 'southeast' ones) larger radii.

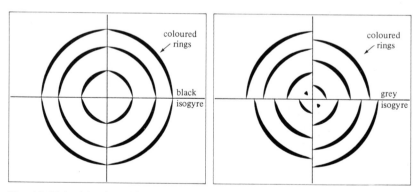

Fig. 46 Uniaxial mineral between crossed linearly polarizing filters.

Fig. 47 Uniaxial mineral between a linear filter and a circular one.

The coloured 'circles' look more like rectangles (plate 79 and fig. 47).

When we put the mineral between two circularly polarizing filters, we see the concentric circles appear again, but the isogyres are absent. This light is no longer characterized by a direction of vibration, but by a sense of rotation.

Uniaxiality occurs in minerals belonging to the tetragonal, hexagonal or trigonal crystal classes. These are (after the cubic ones, which have no double refraction) the classes that have the most symmetrical elements. Crystals with a still lower symmetry are not uniaxial but biaxial. These belong to the remaining crystal classes: the rhombic, monoclinic and triclinic minerals.

68. Biaxial minerals, e.g. aragonite

The phenomena appearing in these kinds of crystal, are not as symmetrical as those in the uniaxial ones, since the crystal structure is also less symmetrical. Instead of concentric circles, a pretzel-shaped figure appears, the two centres forming the optical axes. The distance between the axes varies from mineral to mineral. For the sake of convenience we shall restrict ourselves to the case of crossed linear filters (parallel filters give the complementary image). In this case, and when the direction of transmission of one of the polarizing filters is parallel to an imaginary line between the optical axes, we see the isogyres appear again as a dark cross. One of the isogyres connects the axes and the other crosses it exactly between the axes (fig. 48 and plate 80). Even if only one of the axes is visible (e.g. because the distance between them is too large), the biaxiality is immediately identifiable, because each axis is cut by only one isogyre. The coloured pretzel-shaped figure remains intact when the crystal between the filters is rotated, but the isogyres transform into hyperbolas and no longer cross each other. However, they always keep going through the optical axes (plate 81). The isogyres will fade again when one linear filter is replaced by a circular one, and the coloured figures will shift, as also happened in the case of uniaxial crystals. Between two circular filters, the isogyres are again absent and only the original coloured pretzel-shaped figure will appear.

The rhombic, monoclinic and triclinic minerals are biaxial and consequently show the above-mentioned phenomena. The class of crystal determines the exact way in which the patterns are coloured and so in principle this can be used as a means of identification.

In some cases the measure of biaxiality (the distance between the optical axes) may depend to a large extent on the colour of the light. In this respect it is interesting that the rhombic mineral *brookite*, which in green light behaves as a uniaxial crystal, is biaxial for red and blue light. For these last two colours, however, the 'pretzels' are exactly at right angles to each other

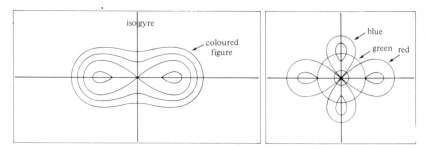

Fig. 48 Biaxial mineral between crossed linearly polarizing filters.

Fig. 49 In brookite the pattern differs from colour to colour.

(fig. 49). In white light a curiously shaped figure appears as a result of the mixture of all these pretzel-like figures of various colours (plate 82). Unfortunately, brookite is a very rare mineral so that in order to observe this phenomenon, we have to visit a mineralogical institute. *Gypsum* is another mineral that has this property, though only a high temperatures (about $90\,^{\circ}C$). However, it is not easy to cleave natural gypsum in the required direction, because it shears much more easily in another direction; in addition, it must not be heated too strongly, since above $120\,^{\circ}C$ it will change into the mineral anhydrite.

69. Optical activity in sugar solutions and in minerals, e.g. quartz

Optical activity (rotation of the polarization plane, fig. 50) in its purest form can be observed in concentrated solutions of sugar and other organic compounds. When such material is placed between linear filters, we can see a colour appear which depends on the positions of the filters with respect to each other and on the concentration of the solution. It is slightly different from the colours that are seen in doubly-refracting materials. Whatever the juxtaposition of the filters may be, there will always be light of some colour that comes through and hence white light can never be completely extinguished. The reason is that the rotation of the plane of polarization

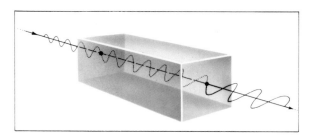

Fig. 50 Passing through optically active material, the plane of vibration of linearly polarized light gradually changes its direction.

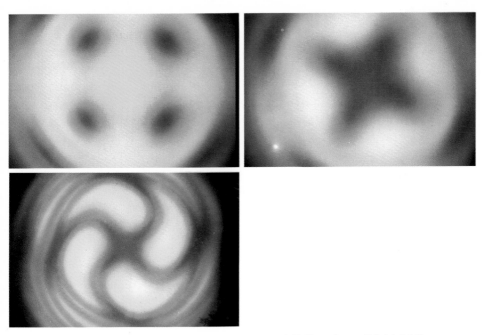

Plates 83–85 *Top:* quartz between crossed (*left*) and parallel (*right*) linear filters. *Below:* Airy's spiral in quartz (§69).

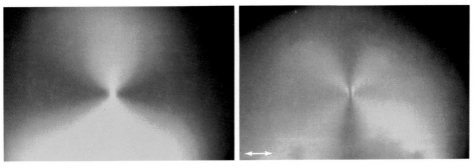

Plates 86–87 *Left:* Brewster's brush in epidote (§70). It bears a strong resemblance to the Haidinger brush (§7). *Right:* with a linear filter in the correct position, idiophanic rings appear in epidote (§70).

depends strongly on the colour (§95). The effect disappears when a circular filter is substituted for one or both of the linear filters, because optical activity can never convert linearly polarized into circularly polarized or vice versa light (unlike double refraction). Consequently, crossed circular filters never transmit light when optically active material has been put between them.

Optically active minerals belonging to the cubic crystal classes show the same effects as solutions of sugar, since double refraction is absent. However, there are hardly any transparent crystals in these classes. In fact, there are only a few minerals in Nature that are optically active, but one of them is present on earth in great quantities – the trigonal mineral *quartz*.

Quartz is both optically active and a uniaxial mineral. The interaction of double refraction and optical activity leads to remarkable interference patterns in polarized light, particularly in the direction of the optical axis. The concentric coloured figures which appear when the quartz is between linear filters have a squarish shape. A coloured area materializes in the centre as a result of the optical activity. When the filters are parallel, this area is 'four-leaved', with 'leaves' at an angle of 45° to the plane of vibration of the light (plates 83–84). This area is always coloured, never white or black as in the case of non-optically active crystals. Farther away from the axis, the optical activity is more and more dominated by double refraction, with the result that the pattern gradually transforms into that of the ordinary uniaxial crystals.

Replacing of one of the linear filters by a circular one causes a great change in the pattern. The figures now join together to produce a spiral, which reaches as far as the centre. In the direction of the optical axis the spiral has two 'wings'. When the second filter is also replaced by a circular one, the same pattern as given by optically inactive crystals will be visible – coloured concentric circles without isogyres – and the optical activity has no influence.

Quartz can have either left-handed or right-handed activity. A curious pattern is formed, when two equally thick sheets of quartz each cut perpendicularly to their optical axis, and which are optically active in opposing directions, are placed upon each other. In the direction of the axis there is no optical activity, so it is dark between crossed linear filters. The isogyres appear, but they are now shaped as a *spiral*! This pattern is called the Airy spiral after its discoverer (plate 85).

When quartz is melted and solidified again, it loses its optical activity, originally the result of a helical arrangement of the *molecules in the mineral* which is lost by melting. It is different with a material like sugar whose activity is the result of a helical arrangement of the *atoms in the molecules*; sugar is optically active, therefore, either when dissolved or in crystalline form (§95). Its optical activity in solid state is difficult to observe, however, as sugar crystals are biaxial and the activity is dominated by strong effects of double refraction. When one wants to see what happens in biaxial optically active crystals, it is better to take a quartz crystal and make it artificially biaxial by means of external stress.

It is worth the trouble to put a sheet of quartz on a sheet of calcite, for example, or to study a sheet of calcite in a sugar solution in order to see what kind of patterns result from these combinations. Coloured spirals can often be seen near the optical axis.

70. Pleochroic minerals, e.g. glaucophane, tourmaline and epidote. Brewster's brush and idiophanic rings

Pleochroic material has the property of showing other colours in different directions of polarization. Such a substance therefore always absorbs light and is usually strongly coloured. The phenomenon is called dichroism for uniaxiality, and trichroism for biaxiality (§84). An example of dichroism is a polarizing filter itself: it is usually clear and greenish in one vibration direction and in another is very dark, deep blue. The latter colour only appears when one filter is crossed with another. Some minerals show the same property, e.g. tourmaline. Here too, the colours are green and deep blue or greenish blue-brown, depending on the variety. The phenomenon is most clearly defined at right angles to the optical axis; it is absent in the direction of the axis. In a trichroic mineral, not two but three colours can appear. Here, which of the colours is visible depends on both the position of the polarizing filter and the side from which one is looking at the mineral. For instance, for glaucophane the combination purple–yellow can be seen on one side of the crystal, yellow–blue on another side and blue–purple on a third side. Without a polarizing filter it also becomes evident that different colours appear on various sides. Cordierite is another example of a trichroic mineral whose three colours differ greatly (see also §12).

The phenomena shown by pleochroic minerals in polarized light are completely different from those of ordinary doubly-refracting crystals. Some of them are visible with only one polarizing filter in front of the mineral or even without filters in unpolarized light, because the material itself functions as a filter.

When we look without filters at the uniaxial mineral tourmaline in the direction of its optical axis, a dark spot appears in the centre of the field of vision. When a linear filter is kept in front of it, a pair of dark brushes will appear perpendicular to the vibration direction of the filter. It is called *Brewster's brush*. A second filter behind it changes the image once more: at right angles to the original brushes, another kind of brush becomes visible, with concentric interference rings around the centre: the *idiophanic rings*. The first-mentioned pair of brushes can be seen in trichroic minerals even with the naked eye! Perhaps the mineral epidote shows it most beautifully (plate 86), but it is also clearly visible in cordierite. One polarizing filter, held in front of or behind the mineral, will suffice to make the idiophanic rings appear in this case (plate 87).

Generally, the phenomena in pleochroic minerals are not nearly so colourful as those in ordinary doubly-refracting minerals. However, the fact that they are so different makes a pleochroic mineral a really interesting object for observation.

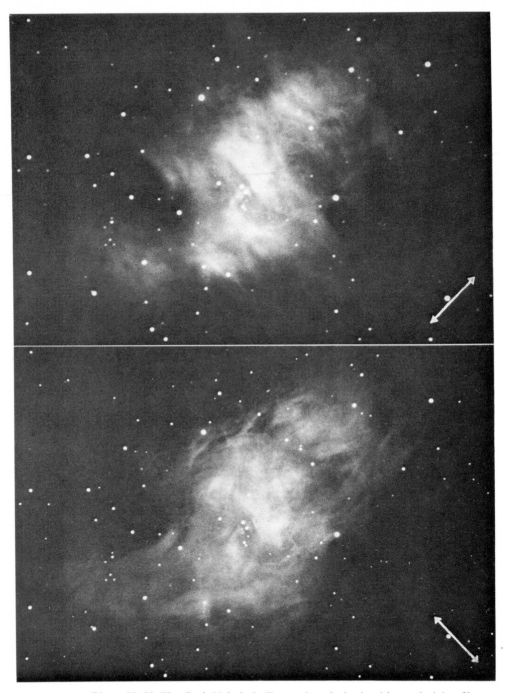

Plates 88–89 The Crab Nebula in Taurus is polarized; with a polarizing filter, its shape changes (§58). This picture has been taken with a very large telescope; with a smaller telescope, the nebula is only visible as a diffuse spot. (Photographs by W. Baade, Mt Palomar.)

Part III

The formation of polarized light in Nature

Introduction

71. General survey

In Nature there are only a few sources of light that emit polarized light. The light from our most important source, the sun, is unpolarized, but nevertheless the quantity of polarized light round about us is considerably. *Indirectly* this light also comes from the sun, but it reaches our eyes by a roundabout way after having been transmitted by objects. It turns out that during this process the originally unpolarized sunlight is often converted into totally or partially polarized light, which is generally *linearly* polarized. *Circularly* polarized light may arise when this linearly polarized light falls subsequently upon a second object. However, by no means all objects are able to convert linearly polarized into circularly polarized light, and moreover, the amount to which this happens, often depends strongly on how the linearly polarized light falls upon the second object. That is why there is much less circularly polarized light around us.

In most cases, the conversion of unpolarized into polarized light, or the conversion of one form of polarized light into another takes place in the same way for all colours. This means that with a polarizing filter all such light can be extinguished to the same degree. When, however, the extent of the conversion of light differs considerably from colour to colour, colour phenomena will be seen when the light is observed with a polarizing filter, but will be absent when such a filter is not used.

In line with the order of importance in which polarized light is formed in Nature, I will first describe the different processes by which unpolarized light is converted into the polarized form. Then, I will deal with the way in which the character of polarized light can change. Finally, I will mention some possible ways in which polarized light can arise directly. While discussing all these mechanisms for producing polarized light, I will systematically refer back to the resulting phenomena in Nature described in the previous part of this book.

The formation of polarized light from unpolarized light

72. Polarization by reflection and by scattering. A general view

Reflection and scattering are the most important ways in which polarized light comes about in Nature. Unpolarized light consists of light waves that vibrate in all directions in turn. When this light falls upon an object, the electrons in the object will start to vibrate in the same directions as the vibrations in the incident light, and subsequently they themselves emit light. The object can be anything: a metal, opaque or transparent material, a minute particle of dust, a single atom or even a single electron. As light waves vibrate perpendicularly to their direction of propagation, the vibrations of the electrons are also normal to it, i.e. their vibrations are only in a plane normal to the beam of light. Looking perpendicularly at this beam, we 'see' all the electrons vibrating on one line only (see also § 5, fig. 10, p. 9). Light emitted by the electrons in this direction, therefore, is also vibrating only in the direction of this line, which is perpendicular to the original beam, and so it is completely linearly polarized. This polarization is tangentially directed with respect to the original source. In another direction, if we look obliquely at the plane in which the electrons are vibrating, and we can observe vibrations with yet another direction. So here, we see only partially polarized light, but the direction of polarization remains tangential.

Small particles like dust, atoms and electrons transmit light in all directions: this is termed *scattering*. On large smooth surfaces like glass, metals etc., the vibrations of the electrons co-operate in such a way that the light is transmitted in one direction only: this is termed *reflection*. Then, too, the polarization is tangential with respect to the source of light and corresponds to a polarization which is horizontal with respect to the reflecting plane. During reflection, a greater or smaller part of the light usually penetrates the material: this is called *refraction*. It is of course evident that a small glass globe reflects light in all directions, and this is also called scattering. Here, the distinction between scattering and reflection is not very sharp.

Usually, the degree of polarization is greatest when scattering or reflection results in a change of direction of about 90° from the original direction of propagation. The angle along which the change of direction has taken place is called the *scattering angle* θ, irrespective whether light is scattered by particles or reflected (refracted) by a surface. Thus θ represents the angular distance between the original source of light

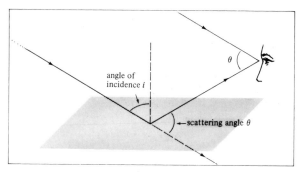

Fig. 51 The scattering angle θ indicates the angular distance between scattered light and the source of light. In the case of external reflection $\theta = 180 - 2\,i$, in which i is the angle of incidence.

and the place in the sky the scattered (reflected) light arrived. The relation between the *angle of incidence i* and the scattering angle θ in the case of external reflection on a smooth surface has been depicted in fig. 51; from this we can infer that the angular distance between the sun and its reflection on quiet water is 90° at a solar elevation of 45°, and 60° at a solar elevation of 30°. Reflection and scattering of light result in characteristic effects of colour and polarization which depend on the size of the scattering particles or the optical properties of the reflecting material. The polarization and colour effects of refraction are essential complementary to those of reflection.

Nearly all the processes that cause conversion of unpolarized into polarized light are related to scattering, reflection or refraction, where refraction may also be followed by absorption of light on its passage through the material. These are the same processes that are responsible for nearly all light we see in Nature, as most light around us has been scattered, reflected or refracted once and sometimes more often.

73. Scattering by small or large particles

Small particles are usually defined as those which are much smaller than the wavelength of light, i.e. much less than about $0.5\,\mu$m. These are for example the gas molecules in our atmosphere, minuscule dust particles and the (even smaller) electrons in outer space. Scattering by these kinds of particles at $\theta = 90°$ results in totally linearly polarized light; the degree of polarization will diminish, when we look at scattered light closer to the source of light or farther away from it (thus with a smaller or larger scattering angle).[1] Light that is scattered backwards or forwards ($\theta = 180°$ or $0°$) has no polarization. No circular polarization is produced at whatever angle one may be looking. The degree of polarization in the case of scattering is the same for all colours, provided that the particles are small. Free electrons scatter white light as white light: this is called *Thomson scattering*. This light is seen as corona light in the neighbourhood of the sun during a total eclipse (§ 19). Atoms and molecules, however, scatter blue light more effectively than red light, because the wavelength of blue light is shorter and, therefore nearer to the size of these particles. This also holds for very small dust or smoke particles, and is

---

[1] The relation between the degree of polarization, P, and the scattering angle, θ, in these cases is $P = \sin^2 \theta / (1 + \cos^2 \theta)$.

called *Rayleigh scattering*. Such scattered light has, therefore, a characteristically blue colour, as does the blue sky for example (§12). However, its polarization is again the same for all colours and it has the same pattern as in the case of Thomson scattering. With a polarizing filter the blue light of the sky can be weakened, particularly at about 90° from the sun, but it retains its blue colour. However, the degree of polarization at 90° from the sun does not amount to 100%, but has a maximum of about 75%. This is the consequence of a number of perturbatory effects (multiple scattering, ground reflections, the presence of dust and molecular anisotropies) that cause the atmosphere to show a pattern of scattering which deviates slightly from the ideal Rayleigh scattering.

In the case of Rayleigh scattering by small particles, the pattern of scattering and polarization is almost independent of the shape and composition of the particles. The intensity of the scattered light is given by the law of Rayleigh and is proportional to λ^{-4}, where λ is the wavelength of light. Larger particles scatter light more strongly because the light is more 'aware' of their presence, and because of this the scattered light becomes less blue. So, if the particle size increases, its scattered light will quickly turn white and completely dominate the blue light of smaller particles in the atmosphere. This can be observed when clouds are forming and is most pronounced at small scattering angles. The polarization of this scattered light, however, continues to show approximately the same pattern. A hazy sky (§13), comet tails (§58), noctilucent clouds (§25), zodiacal light (§60), thin smoke (§22) and scattered light under water (§§55 and 56) all show polarization as a consequence of this kind of scattering. We can also observe this polarization when light is scattered in a bottle of water to which some drops of milk have been added; it is also to be seen in the laboratory when a liquid is close to its critical point – it turns opaque and strongly scatters incident light (this is called critical opalescence).

More complicated effects of polarization, with polarization possibly varying from colour to colour, occur in scattering by particles with dimensions comparable to the wavelength of light. The colour phenomena which appears in nacreous clouds, observed with a polarizing filter, may be related to it (§25).

Finally, when the particles are very large compared to the wavelength of light, the scattering pattern depends greatly on the composition, shape, orientation and optical properties of the particles; it is also very important whether or not the particle is transparent. In order to get a complete picture of the scattering and polarization effects that can occur, we must consider all the reflections, refractions and diffractions of light that can lead to scattering in a certain direction. Such scattering occurs, for example, in water-drops and ice-crystals which each cause different and characteristic patterns of scattering. Rainbows (§27) and haloes (§33) are examples of typical scattering effects by such large particles. It may be preferable to consider this kind of phenomena as the result of reflection and refraction on the surfaces of the particle rather than the result of scattering. But as has already been said, it is not possible to draw a sharp line between these ways of transmitting light in the case of very large particles.

74. External reflection by non-metals

Reflection by smooth non-metallic objects changes unpolarized light into horizontally linearly polarized light; this linear polarization can be total in favourable cases. The conversion takes place in cases of reflection not only by glass, smooth water surfaces, ice etc., but also by shiny wood, smooth stones and grass; in short on all non-metallic objects with a gloss. Since with reflection, as opposed to scattering by small particles, the light is also refracted, the greatest polarization does not happen at a scattering angle of 90° but at a slightly smaller one. The angle where this takes place is given by *Brewster's law* and depends only on the index of refraction of the material, n. Total polarization occurs when $\tan i = n$ or $\tan \frac{1}{2}\theta = 1/n$ (fig. 51). In the case of external reflection n is always greater than 1, so it follows immediately that the scattering angle when the light is 100% polarized, must be less than 90°. When complete polarization is reached, the reflected beam of light is not perpendicular to the incident beam of light but to the *refracted* beam – and from this Brewster's law can be deduced directly from Snell's law of refraction. The index of refraction varies only very slightly from colour to colour, so that the degree of polarization also does not perceptibly depend on colour. When the index of refraction approximates to 1, the polarization pattern is exactly the same as for scattering by small particles; it deviates slightly from this for different indices of refraction. Except at the *Brewster angle* (the angle of incidence for total polarization), the polarization is partial; in the case of normal or grazing reflection, the reflected light is unpolarized. The reflected light is usually much fainter than the transmitted (refracted) light. In principle, there is no difference between reflection on transparent material such as glass or water and reflection on glossy opaque material like stones, skin etc. In both cases, therefore, the gloss can be partially or totally extinguished with a polarization filter. The usefulness of *polaroid sunglasses* is based on this. What remains in the absence of the gloss is the pure colour of the material; substances like glass may become completely transparent when viewed with such a filter (§§51 and 54). The gloss of an object is always white and never takes on the colour of the material, so the world becomes more colourful when we take away the gloss of objects around us with a polarizing filter (§§47–48).

As already stated, the direction of the polarization is *horizontal*. When, therefore, light is reflected by randomly-orientated surfaces or by small spheres, the light thus scattered is *tangentially* polarized, just as in the case of scattering by small particles. We see this light, for instance, in clouds consisting of drops of water or ice-crystals, which whirl around through the atmosphere. The polarization of the light of clouds, therefore, shows approximately the same pattern as the blue sky (§20).

75. Internal reflection by non-metals and total reflection

Here, the laws are the same as those for external reflection. The only difference is that the index of refraction n is now less than 1. For example, when external reflection n is $\frac{3}{2}$ for air–glass then it is $\frac{2}{3}$ for glass–air. This is understandable, when one realizes that the refractive index represents the ratio of the speed of light above and below the reflecting plane. Since the index of refraction in the case of internal reflection is always less than 1, total polarization occurs when the angle of incidence is smaller than 45°. If a phenomenon is generated by internal reflection alone, this implies that maximum polarization is reached for a scattering angle somewhat larger than 90°. In practice, however, it turns out that the maximum polarization of phenomena generated by an internal reflection often is at a quite unexpected place in the sky. The reason for this is that a beam of light must first enter the object before being internally reflected, and refraction then causes an additional change in its direction of propagation. The same happens again when the beam leaves the object. The net result is, that the scattering angle may differ considerably from the one expected from internal reflection alone (fig. 52). This occurs in Nature in the

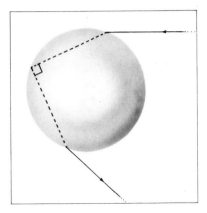

Fig. 52 In the case of phenomena generated by internal reflection, additional refractions may cause the scattering angle to differ considerably from that expected from considering the internal refraction alone. This results in strongly polarized phenomena at quite unexpected places in the sky.

rainbow, which appears at about 140° from the sun, although its light is almost completely polarized by an internal reflection in the raindrops near the Brewster angle (§27). Exactly as in the case of external reflection, the direction of polarization of internally reflected light is horizontal for a flat surface and tangential for randomly oriented surfaces. Consequently one can never infer from the direction of polarization whether it is caused by internal or external reflection. Polarization by internal reflection occurs in Nature not only in the light of the rainbow but also *inter alia* in the light of clouds and some haloes (§§20 and 39).

In the case of a quite oblique incidence, *total reflection* will occur in internal reflection. This means that no light penetrates the surface, so that all incoming light is reflected. This happens when $\sin i > n$, which is only possible if $n < 1$, hence for internal reflection. Totally reflected light is very intense; the totally reflecting surface looks like metal. As entering light of all forms of polarization is completely reflected, no conversion of unpolarized light into polarized light takes place. Total reflection occurs in Nature in the case of formation of some haloes (§39), for example, and in man-made or natural angular transparent objects which then reflect sunlight very strongly (§51).

76. Metallic reflection

When unpolarized light is reflected by a metal, the light is also strongly horizontally polarized, although some textbooks sometimes contradict this. Here, we treat *all* objects with a metallic-type reflection as metallic surfaces, not just metals themselves. It is characteristic of such materials that their reflective capacity is very great and that they are only transparent in very thin layers: the little quantity of light that penetrates the material is very strongly absorbed. It is important for what follows to stress that the nature of the penetrating light is *in all respects* entirely complementary to the reflected light. Gold, for example, has a yellow gloss and, consequently, transmits this colour least well, so that a thin layer of gold transmits a greenish light. The same effect can be observed with sunglasses with a metal coating on the lenses. Moreover, when the reflected light is polarized, the transmitted light is not only contrary in colour but also in polarization. In the case of metallic reflection the polarization is horizontal and thus the transmitted light must be vertically polarized. This is important if we are to understand the polarization effects caused by refraction (§§ 79, 80, 82). The polarization of light in metallic reflection is very considerable, but not so great as in the case of reflection on non-metals. The polarization is maximal when the angle of incidence of the light is rather grazing. This angle, which is comparable to the Brewster angle for non-metals, is closely connected to the *principle angle of incidence* (see § 91). The darker the metal surface, the more (vertical) light can penetrate the metal and, consequently, the stronger is the horizontal polarization of the reflected light. A silver plate, therefore, polarizes the incident light to a smaller extent than stainless steel does (§ 47). The inverse proportionality between albedo (reflection capacity) and polarization also occurs when there is reflection by rough, non-metallic surfaces; it is then called the Umov effect (§ 77).

77. Reflection by rough surfaces

When light is reflected by rough surfaces, the polarization is tangential in this case too, but does not need to be parallel with the surface any more. The degree of polarization usually proves to be less than in reflection by smooth surfaces. The reason is that the light has often been reflected more than once on the various irregularities at the surface before leaving it, and during this process the polarization decreases. This is because the polarized light that has been formed after one reflection against an irregularity may hit a second surface in such a way that its polarization is vertical with respect to the surface. Usually after such a second reflection, the polarization decreases because more horizontal light is added. Very rough surfaces typically do not have a gloss formed by polarized light (§ 75). Many objects in Nature are rough, like snow, sand, stones etc. Although objects such as grass and skin may have a (polarized) white gloss according to the law of reflection for smooth surfaces, the main characteristic remains the diffuse, coloured reflection of the underlying surface (§§ 45–47). The degree of polarization of this light is governed by the *Umov effect* (also called the Umov rule): the *darker* the object, the *stronger* its polarization. So snow, white sand or paper is far less polarized than black sand, asphalt or bricks. According to this rule, the maximum degree of polarization is inversely proportional to the albedo of the material.

The Umov effect can be explained thus. As mentioned above, light that has been reflected once by roughness on a surface, still has a considerable degree of polarization. When it is reflected more often against irregularities at the surface, however, the polarization will disappear to a greater and greater extent. On a dark rough surface, however, multiply reflected light is almost invisible since it has lost too much intensity during all consecutive reflections. That is why in this case we can only see light which is highly polarized. On a bright reflecting surface, on the other hand, we can see a large amount of multiply reflected light (this is the reason why the surface is so bright), but the greater part of the light has lost its polarization completely. The net effect is that in the second case the polarization of the reflected light is less.

A rough surface reflects incoming light in all directions, but most light is still reflected in the direction where we would expect the gloss. When such a surface is exposed to light coming from all sides, as on a sand-plain under an overcast sky, the direction of polarization proves to be horizontal to the surface. In the case of illumination from one source like the sun, the polarization is tangentially directed to the source of light and, therefore, generally forms an angle with the surface. In fact, reflection on a rough surface can be considered as scattering at small rough particles on the surface. The direction of the surface then becomes unimportant and we may expect tangential polarization – which is indeed the case. The only difference between this and scattering by particles in the atmosphere is that on rough surfaces multiply reflected light plays a more important part.

78. Polarized light by scattering and reflection: conclusion

It is evident from the preceding sections that nearly all previously unpolarized light which has been reflected or scattered has become polarized. This polarization is linear and is tangentially (or horizontally) directed. Only when there is total reflection or when light is being scattered (reflected) forwards or backwards (at normal or grazing angles), is there no polarization. So, light phenomena in the direction of the sun or its anti-solar point usually have hardly any polarization, whereas the strongest polarization appears generally at about 90° from the sun. The majority of light visible by day is scattered or reflected sunlight. That is the reason why, in Nature, tangential predominates over radial polarization and horizontal over vertical polarization.

79. Refraction by non-metals

Light falling on an object is split into two beams: reflected and transmitted. In the case of non-metals, the intensity of the reflected beam is usually low: the greater part of the light is refracted and penetrates the material. Yet this transmitted beam is slightly fainter than the entering beam, because a part of it has been reflected. The reflected beam is horizontally polarized. This means that the transmitted light must be vertically polarized: if the reflected and the transmitted beams were to be reunited, the original unpolarized light would have to reappear (fig. 53). As the intensity of the reflected light is usually low, a relatively small part of horizontally polarized light disappears when the beam hits the surface and the polarization of the refracted light is, consequently, rather low. Its polarization is maximal when the angle of incidence is grazing, or when the light leaves the surface at grazing angle. When this happens, the ray has been maximally refracted. The polarization

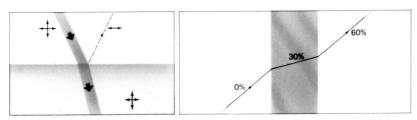

Fig. 53 Refraction by a non-metal. A small part of the incoming light is externally reflected as horizontally polarized light. The remaining part of the light penetrates the material and is, consequently, vertically polarized.

Fig. 54 In the case of one refraction through glass, the maximum degree of polarization is about 30%; refracted twice, it is about 60%.

decreases rapidly when the angle of incidence becomes less oblique. Light that has been refracted once by glass or the surface of water has a maximum degree of polarization of about 30%. When, however, light has been refracted more times, then the polarization is greater: when refracted twice by plate-glass, the degree of polarization of light may be 60% (fig. 54), and when refracted more than twice, the maximum polarization will quickly rise above 90%. A pile of glass sheets, therefore, can be an effective polarizer (§ 7). When we look obliquely at a television screen, the light observed is rather strongly polarized because of the refraction (§ 64). Looking less obliquely, we find that the polarization decreases rapidly. In principle, the luminescent sea should also show this effect, but as the light is usually very faint, its polarization is difficult to determine (§ 61). Apart from these examples, refracted sunlight also is naturally polarized, and this can be observed when we look through a window pane (§ 51) or when a diver looks from under water at the sun (§ 56). The light shining out of windows at night is also polarized (§ 63).

Refraction by randomly oriented surfaces will result in a polarization which has a radial direction with respect to the source of light. In Nature, this phenomenon is shown by some haloes which have been formed by refraction, like the circum-zenithal arc (§ 37). Polarization by refraction can only be expected when the value of the refraction index changes abruptly at a given place. When this change happens gradually, as in the case of mirages (§ 46) or on the surface of the sun (§ 58), there is no polarization.

80. Refraction by metals: emission polarization

The same arguments are valid for refraction both on metals and non-metals. The difference between a non-metal and a metal is that for the latter the reflected light is more intense but its polarization less. This means nevertheless that the small quantity of light that enters the metal must be vertically polarized to a high degree (fig. 55). Here too the polarization is strongest when the angle of incidence is grazing (hence again at a maximum refraction). Normally, we rarely observe this light because of its low intensity and because it is strongly absorbed: metal is only somewhat transparent in very thin layers. But when metal is red-hot, we can see such refracted light! For the red light is formed *in* the metal and on leaving it is refracted and polarized. The polarization is strong and may easily amount to more than 80%. This may happen when the light leaves the metal surface quite obliquely. The surface must be absolutely clean: layers of oxide destroy the polarization. The

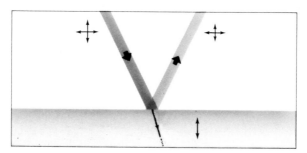

Fig. 55 Refraction by a metal. The reflected light is intense and has a horizontal polarization. The faint refracted light must, therefore, be strongly vertically polarized.

effect is, therefore, clearly visible in the case of silver or platinum and streaming liquid metals like iron (§62). Solid iron does not show this effect because of oxide formation, unless it is *in vacuo*. It is interesting to note that the high degree of polarization of incandescent metals should occur in the case of the ordinary refraction, only if the index of refraction were 4, but the highest index of refraction of transparent matter is only about 3 (for rutile). For glass, such a strong polarization can only be realized after five refractions. The polarization of incandescent materials is also called *emission polarization.*

81. Surface waves

On closer inspection one finds that in the case of total reflection not *all* the light is reflected. A very small part travels as a surface wave (fig. 56). The intensity of this part decreases rapidly, however, as the surface wave proceeds. A surface wave is vertically polarized, just like a sea-wave or the shock waves that travel along the surface of the Earth during an earthquake. Surface waves of light are of a low intensity and generally not perceptible. They prove, however, to contribute significantly to one particular optical phenomenon in the atmosphere – the glory (§42). The polarization of the glory is rather complicated because of the influence of several effects. However, the radial polarization of its rings, which originates from the surface waves, is clearly visible. The appearance of the glory in drops of water is made possible chiefly by the existence of these surface waves. This is also the only phenomenon in Nature in which the polarization is governed by the occurrence of these waves.

82. Optically active metallic surfaces

These are the materials with a metallic gloss that, in the case of reflection, convert unpolarized light directly into circularly polarized light. This is caused by the helical structure of the molecules of which they are composed. Such material may have, for example, the property to transmit right-handed circularly polarized light to some degree, but hardly any left-handed light. In this case also the reflected light has properties which are the opposite of those of transmitted light (§76) and it is therefore almost completely left-handedly circularly polarized. Such light also appears when the reflection is normal to the surface and also when the incident light is linearly or left-handedly polarized. The latter point is remarkable because during

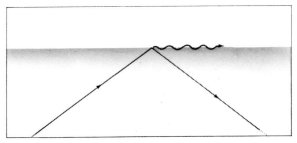

Fig. 56 In the case of total reflection, a very small part of the light travels along the surface as a polarized surface wave.

reflection on other materials the sense of rotation of the circularly polarized light changes (§§88 and 91).

The selective absorption of one form of circularly polarized light is called *circular dichroism*, by analogy to common (linear) dichroism (§84). The phenomenon is closely related to optical activity (§95), which results in a rotation of the plane of vibration of linearly polarized light, as occurs in sugar solution for example. Circularly polarized metallic reflection occurs in some synthetic materials, like liquid crystals. It can be seen in Nature on some beetles, the exoskeletons of which are composed of such optically active material (§50).

The sense of rotation of the reflected light depends on the sense of rotation of the helix of the molecules of which the material consists. This can be left-handed or right-handed in the case of liquid crystals. On the other hand, in living beings, the capacity for producing a given helical molecule is restricted to one sense of rotation. As this sense was fixed at a very early stage in evolution, it is the same for all living organisms (apart from some mutants). Since the exoskeletons of all beetles which are able to convert unpolarized into circularly polarized light consist of the same substance, the sense of rotation of the reflected light is the same for all of them, irrespective of the species, and is left-handed. We see a similar preferential behaviour in the optical activity of sugars: a given biological compound rotates the plane of polarization in a fixed direction. When the same compound is made synthetically, however, it may rotate the plane in the other direction (or may even be neutral, when it consists of a mixture of both types): see §95.

83. Polarization by double refraction

When an unpolarized light beam enters a doubly-refracting (birefringent) material, the transmitted beam is *split* into two beams of light, which proceed with slightly differing paths (fig. 57). It turns out that both beams are completely polarized and have vibration directions perpendicular to each other. Apparently, the index of refraction is not the same for the different directions of polarization in this case. Both beams of light are almost equally bright: the difference between them depends, of course, on the brightness and the polarization of the externally reflected beam of light. Here too, the intensity of the incident unpolarized beam of light must equal

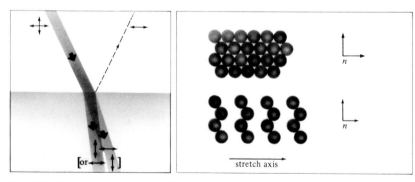

Fig. 57 While passing through doubly-refracting material, unpolarized light is split into two polarized beams.

Fig. 58 When a material is stretched, its index of refraction, n, changes in one direction. Thus it becomes doubly-refracting.

the sum of the intensities of the reflected and refracted beams (§ 76). In the case of double refraction, *four* beams of light are involved, as against *three* in the case of ordinary refraction. Double refraction occurs in a great number of solid materials and minerals: calcite, quartz, stretched plastic, glass with internal stresses, etc. When we put a piece of calcite on a line, we see a double image of the line; both are polarized (plate 59 on p. 86). A piece of quartz can – just as a glass ashtray or a prism – scatter sunlight in a colourful way. We can see coloured spots projected on the wall. These spots change their outward appearance when a polarizing filter is placed in the beam of light or before the crystal; the spots shift, when the filter is rotated (§ 51). Such a shift also occurs in the case of parhelia and other halo phenomena, which result from refraction in (doubly-refracting) ice-crystals (§§ 36 and 37).

The phenomenon is perhaps most easily understood when we consider plastic. In *unstretched* plastic the molecules are arranged randomly and there is no double refraction. In whatever direction the light may vibrate, it always 'sees' the same material and the index of refraction is, therefore, always the same. Such material is called *isotropic*. It becomes *anisotropic*, when the material is stretched in a particular direction: light that vibrates perpendicularly to the direction of stretching, is still 'seeing' the original material, but light vibrating in the direction of stretching, is 'seeing another', because the distance between the molecules is greater (and perhaps also because they have slightly (re)oriented). In the latter direction the material now has another (usually smaller) index of refraction and so the light is refracted in a different way (fig. 58). Consequently, unpolarized light is split into two beams of light that are polarized perpendicularly to each other and follow slightly different paths through the material. The direction of stretching is called the *optical axis*. When we now rotate the material a quarter of a turn so that the light travels parallel to the axis, no splitting occurs, because perpendicular to the beam of light the index of refraction is throughout the same as it was before stretching.

The double refraction obtained by stretching plastic is only weak, but the double refraction of minerals, which are already naturally anisotropic, can be very strong. Calcite is a well-known strongly doubly-refracting mineral and here the difference in the refraction indices is 0.16. Ice is a hundred times less doubly-refracting. Some natural or synthetic minerals such as rutile or sodium nitrate possess an even stronger double refraction than calcite. It depends on the nature of the material whether the index of refraction is higher or lower for vibrations parallel to the optical axis: it is equally possible that the optical axis may be either a *stretch-axis* or a *press-axis*. Of course in the case of minerals, this double refraction is not the consequence of stretching but of anisotropy brought about by the arrangements of the molecules in the material itself. Doubly-refracting materials such as those described above are called *uniaxial*.

It is possible to stretch the material again in another direction, e.g. perpendicular to the optical axis. Such material is called *biaxial*, since there a two directions in it along which entering unpolarized light is not split. These optical axes, however, are not axes of symmetry of the material, as in the case of uniaxial substances. The way in which the double refraction occurs in uniaxial material depends only on the angle between the transmitted beam of light and the optical axis (and is maximum for a light path perpendicular to it). In the case of biaxial material this splitting has a more complicated relationship with the path of light through the crystal. Biaxiality is the highest anisotropy which can exist in material: an additional stretching changes only the way in which it is biaxial. So, *triaxial* materials do not exist.

If only the splitting of unpolarized light into two polarized beams is considered, it is often not important to distinguish between uniaxiality and biaxiality. The difference becomes essential, however, for the description of coloured interference patterns in minerals in polarized light (§§67, 68, 93 and 94).

In 1690, polarization was discovered by Huygens, who studied the double refraction of calcite. One piece of calcite splits a beam of light into two; a second piece splits both beams once again. When we turn these pieces with respect to each other, not only will the four beams shift their positions but pairs will be alternately extinguished. Then, two beams instead of four are seen: the second piece of calcite functions as a kind of polarizing filter. It is worth repeating this historical experiment. It will also become evident that phenomena of *external* reflections on doubly-refracting material hardly differ from those on isotropic material (§74).

84. Polarization by selective absorption (dichroism and trichroism)

This occurs, for instance, in polarizing filters (§§7 and 70). Here, light vibrating in any particular direction is extinguished, because it is more strongly absorbed than light vibrating perpendicular to it. Therefore, the latter component remains. Consequently, unpolarized light is gradually being converted into linearly polarized light (§5). This phenomenon is caused (just as in the case of double refraction) by a difference in structure of the material in different directions. In this case, both the index of refraction and the coefficient of absorption of the material now show differences in various vibration directions, with the above-mentioned result. Not only polarizing filters but also a number of minerals, e.g. tourmaline, have this property which is also called *dichroism*. In co-operation with internal reflections, this effect can sometimes result in surprising changes of colour. This is one of the reasons why tourmaline is sometimes used as an ornamental stone. Many materials exhibit dichroism for some colours only, so that, for example, red light is virtually unpolarized yet blue light is almost totally polarized. Then, with a polarizing filter, we can observe a colour change. Sheet polarizers also have this property, as they do not polarize all colours to the same degree: indeed even crossed filters usually transmit a faint blue light (§70). Other types of polarizing filters are even more transparent in the crossed position, at least for one particular colour (§7).

An anisotropic absorbing material can be biaxial too. In that case, the colour of the polarized light depends on the direction from which one looks through the material. This phenomenon is called *trichroism* and occurs strongly in epidote, glaucophane and cordierite. The collective noun for dichroism and trichroism is *pleochroism*.

Light reflected by a dichroic material with a metallic gloss can be expected to be strongly polarized. The explanation of this is analogous to that of circularly polarized light that is reflected from optically active metallic surfaces (§82): one vibration direction of the light enters the material to a greater or lesser extent (and is strongly absorbed), hence the waves which vibrate in the opposite direction are the most strongly reflected.

Finally, I do not want to leave a remarkable case of dichroism undiscussed: what happens when light passes through oriented needles, which themselves need not be dichroic. The orientation of these oblong objects causes anisotropy and therefore dichroism. This means that light, passing through such needles, becomes polarized, the degree of polarization often depending on the wavelength (= colour) of the light. Polarizing filters can be made in this way. This mechanism proves, however, to be of very little importance for the production of naturally polarized light. On the other hand, it is interesting that it is responsible for the polarization of starlight: on its long way towards us this light passes interstellar dust particles that have been oriented by the magnetic field of our galaxy. This polarization, however, is very low and amounts at the most to only a few percent and it cannot be seen with the aid of simple devices. The effect, nevertheless, is worth mentioning, because much research has been carried out on it by astronomers (§58).

The alteration of the state of polarization by reflection and other mechanisms

85. Introduction

Polarized light that is reflected, scattered or refracted, or passes through a medium, usually changes its character. This change can be an increase or a decrease of the degree of polarization, a modification in the direction of vibration in the case of linearly polarized light, a reversion of the sense of rotation in the case of circularly polarized light or a conversion of part of the linearly polarized to circularly polarized light and vice versa. A number of these changes are reversible: when for example circularly polarized light is converted by a particular process into linearly polarized light with a certain orientation of the plane of polarization, the converse is usually also possible by the same means.

Conversion of one kind of polarized light into another is important in Nature. For instance, nearly all circularly polarized light around us has arisen from linearly polarized light. It is interesting that, in a number of these conversions, the degree to which this occurs depends greatly on the colour of the light. In these cases beautiful colours can be observed with a polarizing filter which are not visible to the naked eye.

In this chapter, I shall describe the processes that lead to changes in the kind of polarization and those which are important for the formation of polarized light in Nature. The initial sections describe conversions which are almost independent of colour. Subsequently, I deal with processes in which the conversion depends very highly on colour.

86. Scattering of polarized light and multiple scattering

If particles scatter light originating from a polarized instead of an unpolarized source, the main difference is that the intensity distribution of the scattered light may lose its cylindrical symmetry around the source. For example, if the sun were to emit horizontally polarized light (which is not the case), the observer would see a very dark sky at 90° from the sun if the sun is near the horizon. The sky exactly over him, however, will be bright. This is exactly the picture that we see, when we look through a horizontally directed polarizing filter at the blue sky lit by the unpolarized sun at a low elevation. Of course, the former situation does not occur in practice. Yet scattering of a polarized source of light can occur at twilight, when the source of light is the polarized blue sky, which displays over almost the entire sky a partial polarization, directed perpendicularly to the set sun (hence, north–south, when the sun is below the western or eastern horizon (§ 14)). During twilight, the scattering pattern of this light under water must therefore display some minima because of the polarization of the light source (§ 56).

Sunlight may reach us indirectly, via particles, by single scattering or by multiple scattering. The latter can even be the dominating process. This happens chiefly when the sun is invisible, i.e. under an overcast sky. In that case particles are lit from all sides by light that was scattered earlier. If the intensity of light from all directions is equal, no polarization will arise as a result of additional scattering on a particle, which can readily be concluded from considerations given in § 72. However, even under a heavily overcast sky a particle receives more light from the zenith (where the cloud screen is thinnest) than from the horizon, so that there is indeed a preferential direction in the diffuse incident light. Because of this, the multiply scattered light is, at least near the horizon, still somewhat polarized. The direction is once again tangential with respect to the mean source of light, so in this case it is horizontal (§ 24).

Multiple scattering also occurs in a clear sky. In that case, the blue light of the sky is once again scattered by the particles in the air and consequently becomes an even deeper blue colour (§ 73). As the singly scattered light comes more or less from all sides, the polarization of the deep blue multiply scattered light is comparatively weak, the more so if yet more scatterings are involved. In broad daylight, multiply scattered light is completely dominated by singly scattered light. But if the latter is maximally extinguished by a polarizing filter, the multiply scattered light becomes visible and the colour of the sky turns a deeper blue (§§ 12 and 13).

By twilight, multiple scattering may yield the dominant contribution to the light in the region in the sky opposite to the sun, so that the polarization of the multiply scattered light becomes visible too. This polarization is vertical and arises because particles still receive less light from the blue sky above them because here the luminous atmosphere is thinnest. Of course it receives hardly any light from below, as it is already dark there. Hence, most light comes from the horizon. This light is already vertically polarized, since it vibrates tangentially with respect to the sun. When this partially polarized singly scattered light is once again scattered, this secondary scattered light is also vertically polarized because (a) the mean position of its source of light is on the horizon and (b) this source of light is on average vertically polarized. Because of these effects the area in the sky opposite to the sun still shows a clear vertical polarization (§ 14), where we should expect from a single scattering a faint horizontal (= tangential) direction of polarization. The clouds in this part of the sky also display the same anomalous direction of polarization (§ 23).

During a total solar eclipse, multiple scattering is much more important for the aspect of the sky: here the reason that the sky is not completely dark, is that the light from outside the zone of totality can penetrate via multiple scattering. The light of the solar corona is only faint and is of minor importance to the illumination. Since multiple scattering dominates completely in this case and the mean source of light is on the horizon, vertical polarization shows up all over the sky during totality (§ 19).

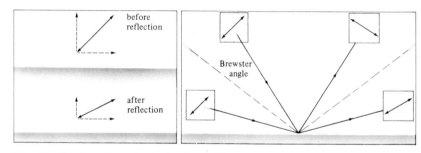

Fig. 59 The direction of polarization of linearly polarized light becomes more horizontal after a reflection.

Fig. 60 The plane of polarization is mirrored when the angle of reflection is steeper than the Brewster angle. This does not happen in the case of reflection at a more grazing angle.

87. Reflection of linearly polarized light by smooth non-metals

Vertically polarized light is considerably weakened when reflected; at the Brewster angle the intensity of the reflected light even falls to zero. Horizontally polarized light is much less weakened than vertically polarized light. In general, one can decompose light with an intermediate polarization direction, e.g. light whose plane of polarization forms an angle of 45° with the surface, into a horizontally and a vertically polarized component which vibrate coherently (§1). As the horizontal component is the least weakened in the case of reflection, the plane of polarization of the sum of the components will become *more horizontal* (fig. 59). At the Brewster angle the plane of polarization of the reflected light is, of course, exactly horizontal.

Except for this rotation, nothing happens to the plane of polarization when light has a rather grazing incidence. This holds generally, when the angle of incidence is larger than the Brewster angle. However, when the angle of incidence is smaller than the Brewster angle and the incidence is therefore steeper, the plane of polarization also becomes to be *mirrored* (fig. 60). The reflection of a polarizing filter can thus have a polarization opposite to that of the filter itself! This has been demonstrated in plate 52 on p. 83 with a common mirror, which in this respect has the same property as a smooth non-metal (§91). The contrariety of the polarization of the image is of course best seen, when the plane of polarization of the filter is at an angle of about 45° to the reflecting surface.

This mirroring is caused by an additional phase-shift of 180° in vertically polarized light on reflection, which does not occur for horizontal light. At an angle of incidence greater than the Brewster angle, this extra phase-shift is absent (fig. 61). The resultant effect can, in principle, be studied by viewing the reflection of a linearly polarizing filter through another one, while varying the position of the first filter and hence its angle of incidence. Then, depending on its position, the direction of polarization of the reflection is mirrored. The test is, however, more convincing with two circular filters, where for the same reason the sense of rotation of the reflected light is reversed at the Brewster angle (see §88). In Nature such mirrorings of the polarization plane are seen when polarized light falls upon a smooth water (or glass) surface (see also §§29 and 48).

The above-mentioned phase-shift can also manifest itself in a completely different way. This happens when there is interference of two light waves, of which one has been reflected at an angle larger and the other at an angle smaller than the

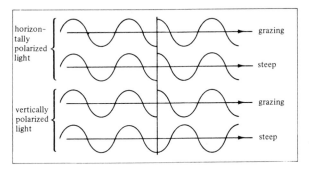

Fig. 61 Phase-shift in the case of external reflection. When the angle of incidence is steeper than the Brewster's angle, vertically polarized light is subjected to an additional phase-shift of 180° with respect to the other possibilities.

Brewster angle. Which colours appear in the case of such interference depends on the difference in the *optical path-lengths* of light of the interfering beams (this is the length of the path expressed in the number of waves of light). As can be established with a polarizing filter, the interference patterns differ in various polarization directions; the reason is, that because of the extra phase-shift, the optical path for vertically polarized interfering light beams differs by half a wavelength (180°) from that of the horizontally polarized interfering light beams. This occurs in Nature in supernumerary fog-bows, which are formed by such interference. As a result of this, the supernumerary bows are shifted when the light is maximally extinguished with a filter (§ 31).

Finally, what happens when the reflecting light is not totally but is only partially polarized? In that case the effects of reflection of the unpolarized and polarized parts of light can be summed: the unpolarized component is partially converted into horizontally polarized light and the polarized part rotates its plane of vibration towards the horizontal direction, while in addition this plane may or may not be mirrored. The net effect is an increase in the degree of polarization and a stronger rotation of the polarization plane towards the horizontal direction than is the case when the incident light is totally polarized.

88. Reflection of circularly polarized light by smooth non-metals

This light becomes elliptical because the 'light circle' is flattened: the vertical part of the circle is less reflected. In other words: a part of the circularly polarized light is converted into linearly polarized light. This can be understood by considering circularly polarized light as the sum of a vertical and a horizontal linearly polarized components, which are shifted 90° in phase (§ 2). The vertical component is reflected to a lesser extent so that the horizontal component will dominate. The direction of polarization of the linearly polarized light, thus added to the circularly polarized one, is, of course, horizontal. In this respect, circularly polarized light behaves exactly like unpolarized light when reflected: a part of the unpolarized/circularly polarized light is converted into horizontal linearly polarized light (§ 74).

Just as stated in § 87, the vertical component is subjected to an additional phase-shift of 180° when the angle of incidence is smaller than the Brewster angle. This does not happen in the case of horizontally polarized light. This means that for

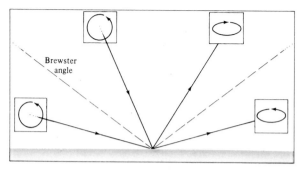

Fig. 62 Circularly polarized light is partially converted into linearly polarized light in the case of reflection. When the angle of incidence is steeper than the Brewster angle, the sense of rotation is reversed.

circularly polarized light, at a rather steep angle of incidence, the sense of rotation is reversed. This does not happen when the incidence is more grazing (fig. 62). When circularly polarized light is reflected at a steep angle by a surface, it can no longer pass through the filter from which it originally came (see also §91 and plate 74 on p. 114). When a circular filter is placed before an object (e.g. before luminous numerals, §64), this effect is used to suppress inconvenient reflections. Then a luminous object will be seen to emit linearly polarized light, as a result of the special construction of the circular filters (§7). The reversion of the sense of rotation, as happens in the case of a steeper angle of incidence and not in the case of more grazing incidence, can be studied graphically by viewing the reflection of a circular filter with a second filter against a windowpane, as described in §87.

89. Reflection of linearly polarized light by rough surfaces

A substantial part of the light is *depolarized* by the multiple reflections to which it may be subjected on rough parts, in a way similar to that described in §77. This depolarization, which means a decrease in the degree of polarization, is strongest for bright surfaces on which light has been reflected many times. This is, of course, in agreement with the Umov effect (§77). However, some polarization usually remains. Therefore, the reflection of the vertically polarized area of the blue sky after sunset on a white sand plain, for instance, is vertically polarized (§45). On a dark asphalt road, this polarization is even stronger. When the sun has risen above the horizon, this vertical polarization is in principle not visible, since the reflection of the bright sunlight dominates completely and therefore determines the direction of polarization. So, the situation changes at sunset. However, at about 90° from the sun, this does not result in a sudden change in the direction of polarization, because here the vertical direction happens to correspond to tangential with respect to the sun. This is the same polarization direction to be expected from direct illumination by the sun.

Reflection of linearly polarized light by rough surfaces can be seen in Nature at twilight on plains or on rough water surfaces (§§45 and 54). By day, it is also visible in the shadow of objects (§46). The rough lunar surface also acts as a depolarizer. This is the reason that the Earth-shine of the moon is virtually unpolarized, though the original source is polarized light coming from our own planet (§58).

90. Total reflection of polarized light

If total reflection takes place, both horizontally and vertically polarized light are completely reflected. For vertical light, this reflection is accompanied by an extra phase-shift, just as in the case of reflection on smooth surfaces with an angle of incidence smaller than the Brewster angle (§87). Now, however, the phase-shift is not 180° but varies according to the angle of incidence and the index of refraction of the material. The phase-shift is greatest at one specific angle of incidence, and the higher the index of refraction the greater the maximum phase-shift that can occur. As a consequence of this shift, part of the incident circularly polarized light is converted during total reflection into linearly polarized light, and vice versa. The total *quantity* of polarized light, however, remains the same. Linearly polarized light is only converted entirely into circularly polarized light if the phase-shift amounts to 90° and the plane of polarization of the incident light is at an angle of 45° to the reflecting surface. Only for a few materials, like diamond or rutile, is the index of refraction so high that this shift can be attained with one total reflection. For most materials, at least two reflections are required for the complete conversion of linearly polarized into circularly polarized light. Yet sometimes one can observe that light has become circularly polarized as a result of total reflections. Sunlight that has passed through an angular transparent object (e.g. a glass ashtray) can have been subjected to so many total and non-total reflections that indeed some of these combinations result in almost completely circularly polarized light. When such an object is illuminated by the blue polarized sky, the amount of circularly polarized light is usually even higher (§§11 and 51). It also turns out that under water, circularly polarized light can be produced by total reflection (§56).

When linearly polarized light has been totally or partially converted into circularly polarized light by a total reflection, the sense of rotation of this light depends on the angle formed by the plane of polarization of the incident light with the reflecting surface. When this plane, seen from the position of the observer, slants upwards from left to right, right-handed light will be formed and vice versa (at least if the object lies on the ground; see fig. 63). Since in the case of total reflection the angle of incidence is always larger than the Brewster angle, the remaining linearly polarized part of the reflected light has the same plane of vibration as the incident beam had before the reflection and is therefore not mirrored (§87).

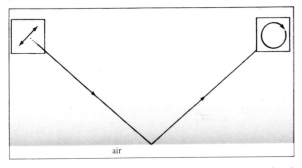

Fig. 63 Conversion of linearly polarized light into circularly polarized light by total reflection.

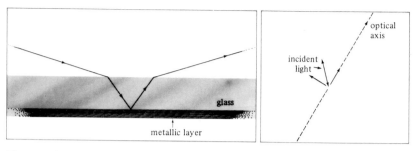

Fig. 64 The angle of incidence on the metal layer of a mirror is always rather steep; at almost any angle of incidence the plane of polarization is mirrored.

Fig. 65 On passing through a doubly-refracting material, linearly polarized light is split into two vibrations, perpendicular to each other; one of them vibrates parallel to the optical axis. Each vibration is subject to a different index of refraction.

91. Reflection of polarized light by metallic surfaces

In some respects such a reflection can be compared to that on a non-metal. Here too, no mirroring of the polarization plane occurs when the angle of incidence is very oblique but it happens when the angle is more vertical, again as a result of an extra phase-shift of the vertically polarized component. In the case of metals, however, the phase difference between the horizontal and vertical components *gradually* increases from 0° to 180°, when the angle of incidence is varied from grazing to normal. This means, that exactly as for total reflection, a part of the original linearly polarized light is converted into circularly polarized light (plates 6–7 on p. 15). At one particular angle of incidence which is called the *principle angle of incidence*, the phase-shift is exactly 90°. Here, therefore, the linearly polarized light can be totally converted into circularly polarized light, when the plane of polarization of the incident light forms the correct angle with the surface. This angle differs from the 45° appropriate to total reflection, because vertically polarized light is reflected less effectively than horizontally polarized light (§ 76); it is called the *principle azimuth*. Complete circular polarization can be produced only when, after reflection, the intensity of the horizontally polarized light is equal to that of the vertically polarized light. Therefore, the angle that the plane of polarization before reflection makes with the surface must be steeper than 45°. The sense of rotation of the circularly polarized light is determined in the same way as in the case of total reflection (fig. 63). Incident circularly polarized light can also, of course, be totally converted into linearly polarized light if reflected at the principle angle of incidence.

There is a strong similarity between the Brewster angle and the principle angle of incidence. In both cases, the phase-shift between horizontal and vertical light becomes larger than 90°, although in the case of the Brewster angle it happens in a jump. The principle angle of incidence is very close to that at which unpolarized light is maximally converted into linearly polarized light (§ 76). When linearly polarized light has been reflected on metals, the plane of vibration of the remaining linearly polarized light rotates to a horizontal direction, as in the case of non-metals (§ 87); when the angle of incidence is steeper than the principle angle of incidence, the plane of polarization is mirrored or the sense of rotation is reversed, because the phase-shift exceeds 90° (§§ 87 and 88). It is, however, true that the principle angle of incidence is generally much larger (consequently more grazing) than the Brewster angle. All the above-mentioned conversions can be seen, when the blue sky is reflected by metals (the chromium on motorcars, § 48).

The same effects occur, in principle, in *mirrors*. The only difference is that the angle of incidence on the metal layer is usually quite steep, since the light has first been refracted by the glass (fig. 64). That is why it may happen that the principle angle of incidence of the metal cannot be reached, even at a grazing incidence on the mirror. The quantity of circularly polarized light that is formed, when linearly polarized light is reflected, is therefore usually not very large, but we do see at almost any angle of incidence how beautifully the plane of polarization is mirrored (plate 52 on p. 83). The reversion of the rotation sense of circularly polarized light can also be demonstrated graphically with a mirror (plate 74 on p. 114).

92. Refraction, double refraction and optical rotation of polarized light. Colour phenomena in polarized light: general description

Refraction of polarized light by isotropic (therefore not doubly-refracting) material does not lead to many new aspects. The plane of polarization is *rotated*, exactly what happens when polarized light is reflected on a non-metal (§ 87). As the vertical component of the light is transmitted more readily than the horizontal part, the rotation takes place in a vertical instead of in a horizontal direction as happens in reflection. This can be observed when polarized light is viewed through a windowpane. However, the rotation is usually not so strong as is the case for reflection, because the difference in transmissibility between both directions of polarization is generally not so large (unpolarized light becomes much less polarized in refraction than in reflection).

Polarized light passing through *doubly-refracting* material, however, behaves differently. The light is split into two oppositely polarized components, each of which follows, with a slightly different speed, its slightly different path through the material (fig. 65). When the double refraction is *strong*, these paths deviate from each other to such a degree that we can clearly see two different beams of light coming out of the material. This is the case with calcite crystals. When the vibration

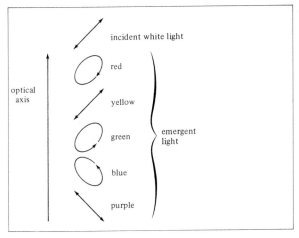

Fig. 66 How the polarization of light is changed after passing through a doubly-refracting material depends on the colour of that light.

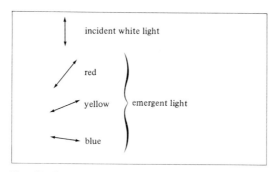

Fig. 67 Conversion of linearly polarized light after passage through optically active material. No circularly polarized light is formed.

direction of the incident light is opposite to that of one of the transmitted beams, this one will disappear and only one refraction will remain; the doubly-refracting crystal seems to have become mono-refracting – in 1808, Malus discovered the polarization of reflections by means of this effect (§10). In the case of other orientations of the polarization plane with respect to the crystal, we do see both differently refracted beams, although in general they will not always be equally bright. Of course, the polarization directions of the emerging beams of light always remain perpendicular to each other (§83).

In the case of a *weak* double refraction, the rays are only split to a small extent so that both polarized beams (originating from the same source) follow almost the same path through the material. Yet it is this situation that leads to splendid phenomena! The *speed* of these waves is slightly different, since the index of refraction for the waves still differs in some degree. Thus, one wave is gradually retarded relative to the other, and this results in a gradual phase-shift between the waves. As a consequence, the polarization of the light beam (i.e. the sum of the vibrations of its components) also changes and after leaving the material this light will have become partially circularly polarized or will vibrate in a different direction (fig. 66). This is the same as what happens when polarized light is totally reflected (§90). A difference is, however, that in total reflection the phase-shift is almost independent of the wavelength of the light, while for double refraction there is a strong dependency on colour: when, after a certain distance through the doubly-refracting material, the difference in path-length between the waves is 0.4μm, this corresponds to one wave with a length of 0.4μm, but with a half-wave of 0.8μm. Hence, the conversion of the original linearly polarized light into light of a different polarization depends strongly on the colour. In the case of incident white polarized light, this results in beautiful interference colours that are invisible without a polarizing filter (plates 62–64 on p. 94). Which colours will become visible at any particular position of a linearly polarizing filter depends on the degree to which one wave is slowed down relative to the other. Of course, similar colours appear in the case of incident circularly polarized light: the processes in either direction are completely reversible. These colours are often more beautiful than the Newton rings (§93). The colour phenomena are fainter when the incident light is only partially polarized. Nothing happens when the incident light in *unpolarized*, exactly as for total reflection of unpolarized light: then, there is essentially no relationship between the phase of waves vibrating perpendicularly to each other, and therefore phase-shifts do not result in colour effects.

Conversion of polarized light by double refraction is also called *chromatic polarization.* It occurs in windscreens of motorcars, in flower-like frost patterns or in hail-stones, minerals, nails, hard plastic with internal stresses or stretched soft plastic and in other transparent materials with mechanical stresses (§§49, 52 and 53).

There are other opportunities to see colour phenomena in polarized light. It occurs in some minerals (quartz) and in sugar solutions. These materials have the property to *rotate* the plane of polarization, and the degree to which it happens depends greatly on the colour of light. Red light is rotated to a much smaller extent than blue light (fig. 67). Hence, colour phenomena appear again in the case of transmission of white, linearly polarized light: as blue light is rotated further than red, we can never extinguish them simultaneously with a second polarizing filter. Because of the strong dependence of the power of rotation on colour, other colours will be seen here than those visible in double refraction. This phenomenon is called *optical rotation*; materials with this capacity are called *optically active*. The colours are only visible when two filters are used (one for polarizing of the entering light and one for observation); and nothing happens in the case of incident unpolarized light. In the case of optical rotation, the direction of vibration of linearly polarized light is altered, but no circularly polarized light will appear, which is the opposite of double refraction. So neither are the colours seen when the incident light is circularly polarized. In the next three sections I shall go further into some aspects of double refraction and optical rotation.

93. Chromatic polarization

As has already been mentioned, the state of polarization of light changes when it passes through a sheet of (faintly) doubly-refracting material. This change depends, however, on the colour of the light. How a particular colour will be polarized depends on how much the waves vibrating perpendicularly and parallel to the optical axis (§83) are slowed down with respect to each other. When this retardation happens to be exactly a quarter of one wavelength, the linearly polarized light will be totally converted into circularly polarized light by a doubly-refracting sheet,

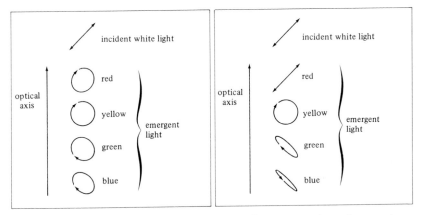

Fig. 68 Conversion of linearly polarized light after passage through a quarter-wave plate (difference in path-length 0.14 μm).

Fig. 69 Conversion of linearly polarized light at a difference in path-length of 0.7 μm.

provided that its optical axis makes the correct angle (45°) with the vibration direction of the incident light. The same happens, of course, when this difference in path-length (and thus the retardation) is $1\frac{1}{4}$ wavelengths, $2\frac{1}{4}$ wavelengths, etc. In these cases, the sheet also converts circularly polarized into totally linearly polarized light.

Red light has a wavelength (λ) of about 0.7μm; yellow, 0.6μm; green, 0.5μm; and blue light 0.45μm. If a doubly-refracting sheet has a difference in path-length of e.g. 0.14μm, then it corresponds to exactly $\frac{1}{4}$ of a wavelength of greenish light of 0.56μm. If, moreover, the optical axis of the sheet forms an angle of 45° with the vibration direction of the incident linearly polarized light, then it is consequently converted totally into circularly polarized light. For the other colours this difference in path-length amounts to 0.20λ (red), 0.28λ (green) and 0.31λ (blue). This means that these colours of light also become chiefly circularly polarized (fig. 68). Such a sheet, which totally converts one colour into circularly polarized light and *almost* totally converts the other colours, is called a *quarter-wave plate*. In the same way, the sheet converts circularly polarized into linearly polarized light. When the sheet is pasted onto a linear filter, a circular filter is formed that functions nearly independently of colour (§ 7).

The situation is different when the difference in path-length amounts, for example, to 0.7μm. It is true that light with a wavelength of 0.56μm is once again totally converted from linearly polarized into circularly polarized light (the difference in path-length now amounts to one complete wave plus one-quarter of a wave), but for the other colours the situation is now different. For instance, for red light of 0.7μm the difference in path-length is now exactly one wavelength and that is why the polarization of this light before and after the sheet will remain exactly the same. Green light of 0.5μm now has a difference in path-length of one wavelength plus 0.4 of a wavelength, and blue light of 0.45μm has a difference in path-length of one plus 0.55 of a wavelength. Apparently the state of the polarization of the emerging light may vary enormously from colour to colour (fig. 69) and when one looks through a rotating polarizing filter one can see other colours. These colours are almost the same as those appearing in, for example, diffraction coronas (§ 43) or in oil-spots on water – the so-called Newtonian colours. You will notice that light subjected to retardation of about half a wavelength, like blue and green in this example, has become almost linearly polarized again, but now the polarization is in the opposite direction (see fig. 69).

A difference in path-length of 0.20 to 0.31 of a wavelength, as produced by a 'quarter-wave plate' for a wavelength range of 0.45 to 0.7μm, is now only present for the much narrower wavelength range of 0.53 to 0.58μm. (The total difference in path-length is, moreover, an additional wavelength, but this does not affect the result.) In this spectral range, where there is little difference in colour, the light behaves, therefore, exactly as it does in a quarter-wave plate for the entire range of the visible spectrum from 0.45 to 0.7μm.

When the material is so thick or so strongly doubly-refracting that the difference in path-length between the interfering beams becomes very large, the colours become invisible again. When this retardation amounts to 10μm, for example, the waves of green light of 0.5μm are slowed down by 20 wavelengths with respect to each other and those of light of 0.53μm, almost the same colour, is slowed 19 wavelengths. The state of the polarization of the transmitted light differs between these wavelengths, but the colours corresponding to these wavelengths are nearly the same (hence, green). The same happens for the other colours. The consequence is that with a polarizing filter we get a mixture of many colours and see almost white

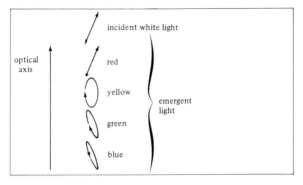

Fig. 70 Conversion of linearly polarized light at a difference in path-length of 0.7 μm, while the vibration plane of the incident light forms an angle other than 45° with the optical axis.

light. Hence, in the case of strong double refraction both the colours and polarization *disappear*. Therefore, such a sheet acts also as a *depolarizer*. We see the most beautiful colours in doubly-refracting materials when the retardation of the waves is not too small and not too large, i.e. when the difference in path-length is about 1–3 wavelengths. Larger or smaller retardations cause the colours to become fainter.

What happens when the optical axis forms angles other than 45° with the direction of vibration? Then the intensity of the component of light vibrating in the direction of the axis is either much greater or much smaller than that of the light vibrating perpendicularly to it. For the sake of convenience we will confine ourselves to the first case; in principle, exactly the same thing happens in the other case. As in the example above, consider a retardation of 0.7 μm in the doubly-refracted beam. After recomposing the light vibrations, it turns out in this case that the light has remained nearly linearly polarized for the majority of the colours, with about the same direction of polarization. Only a *few* colours (green and blue) have a highly altered direction of polarization with respect to the original light, as shown in fig. 70. This means that, with crossed filters, almost all the colours are extinguished, and the few remaining ones offer a beautiful sight against a dark background. From this can be inferred that these colour phenomena are even more brilliant, indeed, than the common Newtonian colours (diffraction coronas, oil-spots etc.), which always appear against a bright background. When, on the other hand, the polarizing filter is parallel to the direction of vibration of the original incident light, almost all colours appear intermixed except for the few colours which we saw with crossed filters. In this case (retardation 0.7 μm), crossed filters show blue light and parallel filters white light with a shade of red: the colour phenomena are completely complementary (white becomes black, and blue becomes red, etc.). From this example, it is evident that with crossed filters the most beautiful colours can be observed, because then no 'white' light is visible, *whatever the position of the optical axis may be*! Consequently, the chromatic polarization of cellophane, plastics, minerals, car windscreens, flower-like frost patterns etc. can be best studied with crossed filters. Moreover, a left-handed and a right-handed circularly polarizing filter may also be used as the crossed filters, and they give rise to the same kinds of phenomena as crossed linear filters do when they form an angle of 45° with the optical axis. These considerations mean, therefore, that generally (in any position of the axis) only a small amount of light with an altered direction of polarization is transmitted. In fact, this is the same as in the case of total reflection: when the plane of vibration forms an angle other than 45° with the reflecting surface, most light will more or less maintain its original direction of polarization, and almost all reflected light will be extinguished with a crossed filter (but of course no colours will appear in this case; see §90).

Finally, when two doubly-refracting sheets overlap, the resulting colour will be different from those of the separate sheets. These colours 'mix', however, in extraordinary ways, and it is quite possible that a combination of red and red, for example, will produce yellow or another colour. The resultant colour depends only on the difference in light path in the combination and this has nothing to do with the original mixing colours. See for example plates 62 and 63 on p. 94.

94. Crystallo-optics (optical mineralogy)

All regular patterns which can be observed in minerals under polarized light usually come under this heading (§§65–70), but in this section I discuss only the non-absorbing and optically inactive minerals in some detail. The phenomena visible in uniaxial minerals are the most easily understood. When we look *exactly* along their optical axis, nothing will happen to the transmitted linearly polarized light (and nor of course to circularly polarized light, which we can consider in the same way). Then, in the case of crossed filters, the light can be totally extinguished. Just alongside this direction, however, there is a slight double refraction, because then we are looking obliquely at the optical axis. The farther away from the axis we look, the more diagonally we look on the axis and the stronger the double refraction will be (fig. 71). As a result, we see more and more colours, as we look farther away from the axis, and coloured *rings* with the optical axis as a central point will appear. Now the colours are at their most beautiful when crossed filters are used. But the angle formed by the axis with the vibration plane of the transmitted polarized light depends on the direction in which we are looking. If from top left to bottom right in fig. 71, this angle is 45°, it is 90° or 0° exactly above the axis or on both sides of it. In this example, the brightest light (with crossed filters) arises top left etc., whereas exactly above or exactly to the left of the axis nothing has happened to the transmitted light, which, therefore, can be extinguished by crossing the filters. By

Fig. 71 Projection of the optical axis, when we look along it in a uniaxial mineral. The farther from the axis we look, the stronger the double refraction will be.

this means, a black cross forms through the coloured circles, the bars of which are called *isogyres*. With parallel filters we see the opposite, a white cross. When using circular instead of linear filters, we see no cross, only coloured circles.

There is a great variety of patterns to be seen in this way. Which pattern appears depends on the applied filters; a combination of a linear and a circular filter produces yet another type (§67). Biaxial minerals show more complicated patterns, which also depend on the angle between the axes and the position of the filters (§68). A combination of double refraction and optical activity produces very remarkable patterns in which the isogyres are deformed to make spirals (§69). Although the explanations for the majority of these kinds of patterns, in principle, follow the same lines as above, it usually turns out that the more detailed description is extremely complicated.

Finally, one may wonder why here only minerals and not stretched plastics, in which generally similar patterns would be expected, are discussed. The reason is that such materials have not usually been stretched in a very regular way so that in general the optical axis meanders through the material. That is why the fine regular patterns, so characteristic of minerals, hardly ever appear in plastics.

95. Optical rotation

An opticaly active material has the capacity to rotate the plane of polarization of transmitted linearly polarized light. When the light has travelled a certain distance through such material, its direction of polarization may have rotated through 90°, for example. Yet the light remains entirely linearly polarized and, consequently, no circularly polarized light is formed. (This is the opposite of what happens in double-refraction phenomena.) The degree to which the rotation takes place, depends very strongly on the colour of the light – when we hold such material between two linearly polarizing filters, we always see colours appear. This is the consequence of the fact that all colours can never be extinguished with a second filter, because each colour has, during its passage through the material, gradually attained another direction of vibration.

How does this curious phenomenon come about? In order to understand it, it is easiest to consider materials which can be stretched (e.g. plastics). Now we must imagine that such a material is not stretched but *twisted*; there is now no stretch-axis but a *torsion axis*. Linearly polarized light moving along this axis 'feels' accordingly no stretch and remains linear. This is different for circularly polarized light. Light rotating *with* the torsion has a slightly different index of refraction from light rotating *against* it, i.e. the light is subject to *circular double refraction*. However, this is generally so weak that the split of incident unpolarized light into circularly polarized light beams is hardly detectable.

Yet this circular double refraction is important. For linearly polarized light can be considered as two opposing circularly polarized beams of light which are vibrating in phase (§2). When linearly polarized light moves through plastic material, one of the component circularly polarized beams of light will gradually slow down with respect to the other. When the emergent beams of light (which now have a phase-difference) are being recomposed, the light has remained linear, but the plane of polarization has been rotated (fig. 72).

The greater the thickness d of the material, the more will be the retardation and so the phase-shift between the component circularly polarized beams, and the greater also the rotation of the plane of polarization. In the case of common double refraction, the phase-shift depends on the number of wavelengths, N, by which the

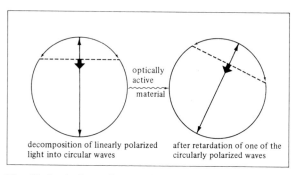

optically
active
material

decomposition of linearly polarized   after retardation of one of the
light into circular waves   circularly polarized waves

Fig. 72 Optical rotation can be explained by circular double refraction.

waves have been retarded with respect to each other. This is given by the expression $N = (n_\| - n_\perp) d/\lambda$. In the case of circular double refraction, $N = (n_L - n_R) d/\lambda$. If appears, however, that the difference in index of refraction between left-handed and right-handed light, $n_L - n_R$, is itself proportional to $1/\lambda$. Thus the phase-shift, and therefore the rotation, is proportional to $1/\lambda^2$ for circular double refraction, while it is only proportional to $1/\lambda$ in the case of common double refraction (since the difference in the index of refraction $n_\| - n_\perp$ in different directions is almost independent of colour). The consequence is that the colours which are visible in the case of optical rotation, differ from the Newtonian colours, because the mixture is different!

Optical rotation occurs in materials that have a helical structure. This can be twisted plastic, for example, but in this material it is not pronounced, as the rotational capacity is small and the torsion axis usually wanders through the material. Here, we generally observe only the common phenomena of double-refraction, which are dominant over those of rotation. Some minerals, like quartz, have a strong internal helical structure. Their optical activity is strong and the splendid phenomenon can be seen by taking a slice from the material and viewing it perpendicularly to the optical axis. The helical structure can, however, also be on molecular level, which is the case with sugar and turpentine. These materials are also active if dissolved, or in liquid or even gaseous state, as Biot found out (§ 10). This is in contrast to quartz, whose helix lies within the crystalline structure and therefore disappears on melting (§ 69). Although the active sugar molecules are randomly oriented in a solution, the rotational capacity is conserved in spite of the fact that some of the molecules are reverse oriented. This is comparable with what is seen in a right-handed thread of a screw, which always remains right-handed even when it is turned upside down.

Optical rotation is seen in its purest form in (sugar) solutions, since no double refraction occurs in dissolved materials and the rotational capacity remains intact (§ 69). In crystals, the colour phenomena of optical rotation must be studied in the vicinity of the optical axis. Again, the finest colours in an optically active material are seen when the thickness of the material or the concentration of the solution is neither too high or too low, exactly as for chromatic polarization. Rotating one of the filters through 90° yields the complementary colours (§ 93).

A given optically active material can be left-handed or right-handed, thus rotating the plane of polarization gradually clockwise or counterclockwise, respectively. So left-handed and right-handed quartz exist in Nature. Which sense of rotation a crystal or molecule will show depends on its helical structure: in principle, both types of handedness have the same probability. If in a solution the quantity of molecules of either type of handedness is equal, the total optical activity of the solution is zero. However, when crystallization takes place, the chance of identical molecules conglomerating is greatest, so that large left-handed and right-handed crystals result from the originally neutral solution. The same happens if optically active crystals are formed from non-active molecules, since during slow crystallization there is a tendency for already extant crystals to grow larger. In Nature, this is the way in which quartz and similar crystals have formed.

In synthetic sugar, equal quantities of left-handed and right-handed molecules are present so that the total optical activity of the solution is zero. Living organisms are also able to produce sugars but in the course of evolution the capacity to produce only one of the two types is left to any individual (§82). So these sugars and many other organic materials also show optical activity when dissolved or in fluid form.

Relatively few minerals are optically active. One exceptional kind ($NaClO_3$) even has this property without being doubly-refracting; the material produces cubic crystals. We can observe in it the same type of phenomena present in sugar solutions. Quartz is the commonest optically active mineral; nearly all other optically active minerals are very rare. Quartz is uniaxial and therefore the optical rotation is only clearly visible near the optical axis. Farther away from it the double-refraction colours prevail. Nonetheless, it is because of this combination that the mineral is one of the most interesting in crystallo-optics (§69).

Sugar crystals show optical activity, but they are biaxial. In this kind of material the optical activity is completely dominated by the double-refraction effects so that it is barely noticeable.

Liquid crystals also may show optical activity (§82). However, it may change when an electric field is applied to it. Nowadays the property is frequently invoked when making numeral displays on digital watches. Here, it is essential that the watch has a polarizing filter in front of its numerals (§64).

Sources of polarized light

96. Introduction

A body which emits polarized light is a source of polarized light. Examples of strongly polarized sources of light in Nature are incandescent metals, the blue sky at 90° from the sun and reflected sunlight. An unpolarized source of light fitted with a polarization filter is, strictly speaking, also such a source.

However, the above-mentioned sources of light are without intrinsic polarization, as their light has become polarized only by the secondary process like scattering, reflection or refraction. But, there are also a few ways in Nature by which polarized light is formed directly. The processes in question are usually events which take place on the atomic level. They sometimes occur in electrical discharges in which light originates from decay of excited atoms. The phenomena I shall deal with include:

Electric discharges and the Zeeman effect
Synchrotron radiation

All these sources emit linearly polarized light. Direct sources of strong circularly polarized light do not exist in Nature.

97. Electric discharges and the Zeeman effect

When atoms are lit up by an electrical discharge, polarization of the emitted light can indeed occur. This is only the case, however, when the atmospheric pressure is very low – less than about one-millionth of an atmosphere. So it does not occur under terrestrial circumstances: sparks, fluorescent light-tube discharges, lightning etc. are, therefore, unpolarized. The polarization of fluorescent lighting exists only because this light is refracted by the glass sheet behind which the discharge takes place (§64). Flames are also unpolarized.

It is otherwise with the aurora, which appears at such great heights in the atmosphere that formation of polarized light becomes possible. However only

160

some of the emission lines (and so only certain colours of the aurora) can be polarized. The high red aurora, which is almost monochromatic, is an example and it can have a degree of polarization of as much as 60%, provided one looks at a distance and from the magnetic zenith. The direction of this polarization is perpendicular to the magnetic field. The lower green or whitish polar lights do not have this polarization, because the atomic transitions that are responsible for them, can never produce polarized light (§ 59). The aurora can be considered as an electric discharge which takes place in the (terrestrial) magnetic field. The polarization that appears is closely related to the Zeeman effect. This is the splitting of an emission line into two polarized components when a magnetic field is applied to the discharge. In its purest form, however, the Zeeman effect does not occur elsewhere in the light around us.

98. Synchrotron radiation

This radiation occurs, when there are very fast moving electrons in a magnetic field. They describe helical orbits, while emitting a bluish white light that is highly linearly polarized. The direction of polarization is perpendicular to the magnetic field (fig. 73). The radiation is very strong at radio wavelengths, and this emission is polarized in the same way as the optical radiation.

Synchrotron radiation does not occur in terrestrial circumstances; it can be produced in laboratories. It turns out, however, that there are some sources of light in the universe that emit this radiation at visible wavelengths. The most famous of these are the Crab Nebula M1 in Taurus and the 'jet' coming from the elliptical galaxy M87 in Virgo. The observation of the former requires a telescope of good quality. The direction of polarization differs from place to place and depends on the local magnetic field. Near the edge of the Crab Nebula the degree of polarization is 70% (§ 58). The polarization of the jet in M87 is much less and can only be observed with highly professional equipment, usually far beyond the range of the amateur.

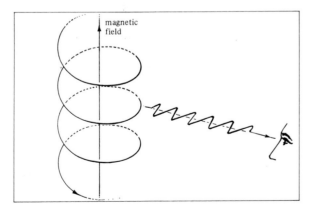

Fig. 73 An electron describes a helical orbit in a magnetic field. When its speed is sufficiently high, the electron emits light that is vertically polarized with respect to the field.

Conclusion and overview

99. Polarized light in Nature: a physical survey

Polarization of the light around us is very common and is often present to a high degree. However, in contrast-to its ability to observe colours, the human eye is scarcely able to perceive polarization of light, so that in this respect Nature remains dull to us. This situation changes, however, when one looks through a polarizing filter because the filter translates the existing polarization into a change in intensity. Since shades of natural intensity are now mixed with shades of intensity originating from polarization, curious light effects appear which would remain concealed without a polarizing filter. Because of the high degree to which polarization often occurs, such a simple and, strictly speaking, insensitive device as a simple polarizing filter suffices to give Nature quite another semblance.

Polarization depends on the previous 'history' of the light, hence on the path traversed from the source of light to the eye. Generally, natural light can best be divided into three classes of decreasing brightness:

1. Direct sources of light (the sun, artificial light, the moon, the stars etc.);
2. Light that reaches us via one object or particle;
3. Light reaching us via several objects or particles.

The *sources* of light in Nature are hardly ever polarized. By night they form the most characteristic part of our world; our eyes are involuntarily drawn to them. By day the sources of light can be so dazzling that we take care not to look straight at them and so they are no longer the most striking part of our world: this function is taken over by the illuminated objects around us.

Light that reaches us *via one object* determines our surroundings by day; it is the brightest light in Nature, next to the direct sources of light. On its

162

way via such an object the originally unpolarized light of the source is now often partially converted into polarized light, which is nearly always linear. Characteristic of the light reaching us via one object is that it has not yet 'forgotten' its origins. The pattern of polarization, therefore, has a cylindrical symmetry around the original source of light, which means, on the one hand, that the direction of polarization must be either radial or tangential with respect to the source of light and, on the other, that the degree of polarization depends only on the (angular) distance to the source of light. This symmetry is only strained, when the object itself shows a very strong anisotropy *qua* structure or *qua* orientation; this can lead to an altered direction of polarization or even, in a rare case, to the formation of circularly polarized light.

By far the most light around us comes about by scattering at small particles or by reflection on rough or smooth surfaces. This leads to *tangential* polarization, which is, therefore, the predominant form of polarization in Nature. *Radially* polarized light remains restricted in Nature to those few isolated light phenomena in which not reflection but refraction, double-refraction or other processes play the essential role in the formation.

Light reaching our eyes via *several particles* or objects is even weaker. Because of all the reflections etc. to which it is subjected on its way from the source towards us, it has, moreover, 'forgotten' its origins, and its polarization pattern will bear no direct relationship to the original source of light. After passage through the various objects, the originally un-polarized light can have been transformed into both linearly and circularly polarized light; and the degree of the polarization, the direction of vibration of the linear part and the sense of rotation of the circular part can always vary depending on the path traversed.

It is the contribution of this type of light in our world that is responsible for nearly all circularly polarized light in Nature and also for colour-dependent polarization effects. This contribution is most explicit by twilight, when the main source of light is already polarized (the blue sky), but in normal circumstances it is so weak in comparison to the light that reaches us via one particle or object that it is hardly able to affect the overall linear polarization with its symmetrical pattern of the world. It merely adds a number of novel possibilities to a world which is dominated by single scattering.

References

1. General references

R. W. Ditchburn, *Light*, Blackie and Son Ltd, London/Glasgow 1963 (optics and polarization).

T. Gehrels (ed.), *Planets, Stars and Nebulae studied with Photopolarimetry*, University of Arizona Press, Tuscon, Arizona 1974 (polarization of many objects).

R. Greenler, *Rainbows, Halos and Glories*, Cambridge University Press, Cambridge 1980 (atmospheric optics).

M. Minnaert, *De natuurkunde van het vrije veld I*, Thieme, Zutphen 1968. English translation: *Light and Colour in Open Air*, Dover 1968 (optics of open air).

G. N. Ramachandran & S. Ramaseshan, *Crystal Optics*, Handbuch der Physik, Vol. 25, ed. S. Flügge, Springer, Berlin 1961 (optical mineralogy).

W. A. Shurcliff, *Polarized Light, Production and Use*, Harvard University Press, Cambridge, Mass. 1966 (polarization).

W. A. Shurcliff & S. S. Ballard, *Polarized Light*, van Nostrand, London 1964 (polarization).

R. A. R. Tricker, *Introduction to Meteorological Optics*, Elsevier, New York; Mills and Boon, London 1970 (atmospheric optics).

S. W. Visser, Die Halo-Erscheinungen, in F. Linke & F. Möller (eds.), *Handbuch der Geophysik*, Bd VIII, Börnträger, Berlin 1940 (atmospheric optics).

R. W. Wood, *Physical Optics*, Macmillan Co., New York 1947 (optics and polarization).

2. Key-references to the sections

When referring to a General Reference, only the author/editor and page are mentioned. The references to Minnaert's book are to the Dutch edition.

Section
§6 K. von Frisch, *Experientia* **5** (1949) 142, Gehrels, p. 472, R. Wehner, *Scient. Am* **235** (July 1976) 106 (animals); D. M. Summers *et al.*, *J. opt. Soc. Am.* **60** (1970) 271 (Haidinger); Minnaert, p. 472 (optical illusion); W. A. Shurcliff, *J. opt. Soc. Am.* **45** (1955) 399 (Haidinger in circular light).

§12 Gehrels, p. 444 (blue sky); J. Walker, *Scient. Am.* **238** (January 1978) 132 (Vikings).

§14 Gehrels, p. 29, N. B. Divari, *Izv. Atm. oc. Phys.* **3** (1967) 289 (negative polarization); G. V. Rozenberg, *Twilight*, Plenum Press, New York 1966 (twilight).

§17 R. Tousey *et al., J. opt. Soc. Am.* **43** (1953) 177 (stars in daylight).

§19 G. E. Shaw, *Appl. Opt.* **14** (1975) 388, S. D. Gedzelman, *Appl. Opt.* **14** (1975) 2881, D. W. Jannink, *Zenit* **8** (1981) 526 (sky during eclipse); J. Pasachoff, *Scient. Am.* **229** (October 1973) 68 (solar corona).

§20 Gehrels, p. 518 (clouds).

§25 Gehrels, p. 514, B. Fogle *et al., Space Sci. Rev.* **6** (1966) 270 (noctilucent clouds); Minnaert, p. 266, J. Hallet *et al., Weather* **22** (1967) 56 (nacreous clouds).

§27 C. M. Botley, *Weather* **25** (1970) 287 (eclipse rainbow).

§29 Minnaert, p. 219 (reflected dew-bow), K. Lenggenhager, *Z. Meteorol.* **26** (1976) 112 (reflection-bow).

§31 G. P. Können *et al., Appl. Opt.* **18** (1979) 1961 (supernumerary fog-bows).

§32 S. Rösch, *Appl. Opt.* **7** (1968) 233 (thirteenth rainbow).

§36 G. P. Können, *J. opt. Soc. Am.* **73** (1983) 1629 (double refraction haloes); K. Lenggenhager, *Meteorol. Rundsch.* **25** (1972) 41 (sub-parhelia of sub-sun); Visser in *Handbuch* p. 1069 (polarization of 22° halo).

§40 Minnaert, p. 255 (double halo).

§41 A. Cornu, *Comt. r. hebd. Séanc. Acad. Sci. Paris* **108** (1889) 429 ($NaNO_3$ halo).

§42 H. M. Nussenzveig, *J. opt. Soc. Am.* **69** (1979) 1068 (glory polarization); Tricker, p. 206 (unpolarized glory).

§43 C. M. Botley, *Weather* **25** (1970) 287 (eclipse corona); Minnaert, p. 262 (iridescent clouds); Tricker, pp. 24 and 146 (corona and heiligenschein).

§44 N. Umov, *Phys. Z.* **6** (1905) 674, Gehrels, p. 384 (Umov effect).

§50 Gehrels, p. 495, G. Robinson, *Molec. Cryst.* **1** (1966) 467 (beetles).

§55 Minnaert, p. 364 (deep oceans).

§56 A. Ivanov in N. G. Jerlov and E. S. Nielsen, *Optical aspects of Oceanography*, Academic Press, New York/London 1974 (polarization under water); Gehrels, p. 434 (circular polarization).

§58 Gehrels, p. 891 (stars); pp. 518 and 381 (planets and moon); p. 814 (comets); p. 371 (artificial satellites); p. 834 (VY CMa); p. 1014 (M1).

§59 R. A. Duncan, *Planet. Sp. Sci.* **1** (1959) 112, J. W. Chamberlain, *The Physics of the Aurora and Airglow*, Academic Press, New York/London 1961, p. 205 (aurora).

§60 J. W. Chamberlain, *The Physics of the Aurora and Airglow*, Academic Press, New York/London 1961, p. 508, Gehrels, p. 780 (airglow); Gehrels, p. 781 (zodiacal light); Gehrels, p. 794 (gegenschein).

§62 Wood, p. 783, O. Sandus, *Appl. Opt.* **4** (1965) 1634 (emission polarization); A. Pflüger, *Annln. Phys.* **7** (1902) 806 (tourmaline).

§64 A. Sobel, *Scient. Am.* **228** (June 1973) 65 (digital watches).

Index

Page numbers in **bold** type indicate the place where the entry is discussed more extensively.